D1521904

Risk Assessment and Environmental Fate Methodologies

Edward J. Calabrese, Ph.D.
Director
Northeast Regional Public
Health Center
University of Massachusetts

Paul T. Kostecki, Ph.D.
Associate Director
Northeast Regional Public
Health Center
University of Massachusetts

Library of Congress Cataloging-in-Publication Data

Risk assessment and environmental fate methodologies / [edited by] Edward J. Calabrese, Paul T. Kostecki.

p. cm.

Includes bibliographical references and index.

ISBN 0-87371-711-2

1. Oil pollution of soils—Environmental aspects. 2. Oil pollution of soils—Environmental aspects—Computer simulation. 3. Oil pollution of soils—Health aspects. 4. Oil pollution of soils—Health aspects—Computer simulation. 5. Pollution—Risk assessment. I. Calabrese, Edward J., 1946— . II. Kostecki, Paul T.

TD879.P4E58 1992

628.5'5—dc20 92-2504

CIP

Direct all inquiries to CRC Press, Inc. 2000 Corporate Blvd., NW
Boca Raton, Florida 33431

PRINTED IN THE UNITED STATES OF AMERICA
1 2 3 4 5 6 7 8 9 0
Printed on acid-free paper

Risk Assessment and Environmental Fate Methodologies

DISCLAIMER

The comments and conclusions expressed in this report represent the general consensus of the respective committees and do not necessarily represent those of individual committee members, sponsoring organizations or companies.

CONTENTS

PREFACE

Soil contamination has been recognized as a significant environmental and public health concern over the past decade. The range of contaminants in soil is now known to be very broad, and includes petroleum constituents, heavy metals, dioxins, pesticides, organic solvents, and other substances of potential environmental and public health concern. The presence of such contaminants in soil is affecting land use throughout the United States, including types of land development and economic expenditures.

Costs associated with cleanup activities have escalated enormously, often to the point of having a major impact on business and residential development. Along with these increased costs of cleanup is the growing recognition that soil contamination may present significant health problems through various exposure pathways (e.g., groundwater contamination, soil ingestion, crop contamination, and localized air pollution).

Recent surveys have indicated important differences at the state and federal levels concerning regulatory approaches for soil contamination. Uncertainty and inconsistency with respect to how contaminated soil is dealt with among federal and state agencies have led to confusion in the private sector concerning what to expect from the regulatory agencies.

Within the context of this background, in November 1987, the International Society for Regulatory Toxicology and Pharmacology (ISRTP) convened a meeting with representatives from the federal and state public health and environmental agencies, the private sector, academia, and the Society at the University of Massachusetts at Amherst, to determine the need for an expert committee to develop a consensus risk assessment methodology for soil contamination.

These representatives were from the University of Massachusetts, United States Environmental Protection Agency (U.S. EPA), Agency for Toxic Substances and Disease Registry (ATSDR), New Hampshire Department of Public Health, Association of State and Territorial Health Officers (ASTHO), Environmental Subcommittee, New Jersey Department of Environmental Protection (NJ

DEP), Electric Power Research Institute (EPRI), DuPont Company, Hercules Incorporated, Procter & Gamble, Inc., Texaco, American Petroleum Institute (API), McLaren-Chemrisk, Inc. Also in attendance were the ISRTP Vice President, secretary, and legal counsel.

It was determined that there was a crucial and immediate need for a consensus risk assessment methodology for determining soil contamination. Furthermore, the representatives concluded that the ISRTP should play a significant leadership role by creating the organizational framework to successfully address critical environmental and public health issues associated with contaminated soils. Subsequently, the Council for the Health and Environmental Safety of Soils (CHESS) was chartered as an entity directly responsible to the ISRTP's Board of Directors.

A Governing Board for CHESS (Table 1) with respected and experienced scientists from the federal government, state departments of public health and environmental protection, the private sector including industry and environmental organizations, and academia was created to establish the goals of CHESS and its mode of operation, and to establish funding policies and directions. The Governing Board, which is designed to oversee and direct the overall directions of CHESS, then created a Technical Council (Table 2) to oversee the development of a peer-reviewed concensus methodology to assess public health risks from contaminated soil by technical committees (Tables 3 and 4) composed of recognized experts in the area of soil contamination and other relevant disciplines. This consensus soil risk assessment methodology would then be made available to federal and state agencies, the private sector, and the scientific community at large.

The Governing Board determined that the goal of CHESS is to provide leadership in soil contamination issues by:

- providing consensus guidelines on analytical techniques, fate of soil contaminants in the environment, risk assessment methodologies, and remediation of contaminated soils
- conducting scientific evaluation, making analyses, and providing recommendations for a course of action

- exchanging technical information, especially in the peer-reviewed and open literature
- providing education and training functions, especially for professionals at the state level
- encouraging dialog among affected groups

It should be noted that much initial debate centered on whether CHESS should provide specific recommended cleanup levels, or alternatively, a process by which more effective levels (i.e., cleanup standards) would likely to be derived at state levels. The final decision was that soil-related risk assessments are so strongly affected by site-specific factors such as types of soil, depth to groundwater, annual precipitation variation, proximity of biological receptors, land use options, etc., that actual numerical values would not be advisable. However, it would be more useful to provide a user-friendly decision-tree framework that could be used by professionals at the state and local levels for application to nearly all types of soil contamination sites. While this decision framework may not provide a quick fix for a specific site, it will provide a more defensible risk assessment framework with high adaptability to nearly all possible sites to be encountered. Within this context, the Governing Board believes that the approach is most suited to state needs while not putting the states on a possible collision course with a specific recommendation by CHESS.

In order to provide financial support for Council activities, funding was solicited from various organizations including federal and state agencies, foundations, and the private sector. While a listing of all financial supporters to date is provided in Table 5, the principal supporters have been U.S. federal governmental agencies including the Agency for Toxic Substances and Disease Registry (ATSDR) and the U.S. EPA. The financial support was principally designed to pay for the management of the Council activities through a contract to the University of Massachusetts. Council members offered their respective services voluntarily to CHESS without remuneration.

The Council set forth to assess the most effective approach for developing practical solutions concerning contaminated soils. To this end, the Council determined that the first initiative should focus on a specific soil contamination issue and that the approach should be a product that is constructed within an expert decision framework. The rationale for focusing on a single type of soil contamination problem was that, although it may be possible to develop a generic methodology applicable to all soil contamination instances, it is not feasible within an acceptable time frame. Therefore, the Council felt that focusing on a specific area would help to ensure the development of a usable product for states on a specific area of concern and facilitate the development of a model approach that could be applied to other pollutant classes.

Discussions concerning the first area of concentration revolved around the collective experiences of Council and Governing Board members. In addition, input was solicited from various state agencies regarding the significance of soil contamination problems in terms of dollars, manpower, and time. Specific information was provided via videotape from representatives from the Orange County Health Care Agency, Santa Ana, California; the Arizona Department of Environmental Quality, Phoenix, Arizona; and the Massachusetts Department of Environmental Quality Engineering, Worcester, Massachusetts. In addition, the results from an ongoing national survey being conducted at the University of Massachusetts assessing the significance of petroleum soil contamination were provided to the Council.

States unanimously agreed that the most pressing soil contamination problem they face is petroleum contamination. Agencies indicated that anywhere from 60 to 90% of their resources in the soil contaminant arena were being directed to petroleum contaminated soils. The Council also felt that the ubiquitous nature of petroleum contamination has direct and immediate public health implications when the health hazard of the constituents of petroleum products and the exposure potential for large segments of the population are considered. The Council, therefore, overwhelmingly voted to have CHESS' first efforts directed to this area.

The Council has charged the committees with evaluating all relevant methodologies and approaches (Tables 6 and 7) and providing recommendations as to their use in the development of a comprehensive expert decision methodology with direct application to petroleum contaminated soils. The process was designed to provide the user community (i.e., state and federal agency, industry, and consulting personnel) a common ground for determining the use of methodologies and approaches which could be applied to petroleum contaminated soils.

In the intervening two years, intensive activities of the members of the Analysis and Environmental Fate and the Environment and Health committees produced this initial CHESS product. Finally, the committee Chairs synthesized the reviews of their committee members and recommended the most appropriate model/approach for the Council to adopt. These findings are provided in the Executive Summary of this document.

This book is an edited summary of the reviewers' comments which were compiled and synthesized to reflect accurately the tenor of the individual reviews *including disagreements*. The committees' efforts represent enormous time commitments and dedication from each member. Numerous methodologies and enormous amounts of documentation were reviewed. Since some methodologies included both environmental fate and health components, they were reviewed by both committees. The entire review effort is represented by the following points:

- twenty committee members and advisors to the committee reviewed 16 state-of-the-art methodologies
- the review material and supportive documentation approached 10,000 pages
- over 300 pages of review comments were produced and synthesized into the present report
- it is estimated that the present report represents more than 2,000 man-hours

Each chapter contains an objective description of an individual model/approach with referenced documentation to direct the user to a more in-depth description. Information regarding the availability of each approach is also provided. The evaluation sections provide the reader guidance through a discussion of each approach as to its basis in science, applicability, ability to address multiple environnmental media, input data requirements, and general strengths and weaknesses.

Table 1

Original CHESS Governing Board

Chair:

John Frawley, Ph.D.
International Society of Regulatory Toxicology and Pharmacology

Members:

Jelleff Carr, Ph.D.
International Society of Regulatory Toxicology and Pharmacology

John Doull, Ph.D., M.D.
Department of Pharmacology, Toxicology, and Therapeutics
University of Kansas Medical Center

Robert Drew, Ph.D.
Health and Environmental Science Department
American Petroleum Institute

William Farland, Ph.D.
U.S. Environmental Protection Agency

Peter Galbraith, D.M.D.
Connecticut Department of Health Services

Barry L. Johnson, Ph.D.
Agency of Toxicology, Substance and Disease Registry
Centers for Disease Control

Donald R. Lesh, Ph.D.
Global Tomorrow Coalition

David G. Loveland
Director—Natural Resources
League of Women Voters

Anthony Marcil
World Environment Center

Gordon Newell, Ph.D.
Electric Power Research Institute

James Solyst
National Governor's Association

Robert K. Tucker, Ph.D.
New Jersey Department of Environmental Protection

Note that as of publication, composition of Board will have changed.

Table 2
Original CHESS Council

Director:

Edward Calabrese, Ph.D.
Director
Northeast Regional Public Health Center
University of Massachusetts

Managing Director:

Paul Kostecki, Ph.D.
Associate Director
Northeast Regional Public Health Center
University of Massachusetts

Members:

James Dragun, Ph.D.
The Dragun Corporation

Allen Hatheway, Ph.D.
School of Mines and Metallurgy
Department of Geological Engineering
University of Missouri — Rolla

Dorothy Keech, Ph.D.
Chevron Oil Research Co.

Renate Kimbrough, M.D.
U.S. Environmental Protection Agency

Robert Menzer, Ph.D.
U.S. Environmental Protection Agency

Warner North, Ph.D.
Decision Focus

Dennis Pastenbach, Ph.D.
McLaren Hart

Brian Strohm, Ph.D.
Department of Health and Welfare
Division of Public Services

Note that as of publication, composition of Council will have changed.

Table 3
Environment and Health Committee Members

Chair:

Dennis Paustenbach, Ph.D.
Chairman
McLaren Hart

David Rosenblatt, Ph.D.
Co-Chairman
Private Consultant

Barbara Beck, Ph.D.
Co-Chairman
Gradient Corporation

Members:

Cynthia Harris, Ph.D.
ATSDR — Office of Health Assessment and Consulting

Renate Kimbrough, M.D.
U.S. Environmental Protection Agency

David Layton, Ph.D.
Lawrence Livermore National Laboratory

Richard McKee, Ph.D.
Exxon Environmental Science Corporation

Randy Roth, Ph.D.
ARCO

John Schaum
U.S. Environmental Protection Agency Office of Health and Assessment

Jeffrey Wong, Ph.D.
California Department of Health Toxic Substance Control Division

Note that as of publication, composition of Members will have changed.

Table 4

Analysis and Environmental Fate Committee Members

Chair:

James Dragun, Ph.D.
The Dragun Corporation

Members:

Bruce Bauman, Ph.D.
American Petroleum Institute

Dwayne Conrad
Texaco Research

Donald Mackay, Ph.D.
Department of Chemical Engineering and Applied Chemistry
University of Toronto

Marcos Bonazountas, Ph.D.
National Technical University of Athens

Thomas Potter
Agricultural Experimental Station
University of Massachusetts

Table 5
CHESS Financial Supporters

Agency of Toxic Substances and Disease Registry, Atlanta, GA

Ashland Oil Inc., Ashland, KY

Chevron Inc., Richmond, CA

Eastman Kodak Company, Kingsport, TN

Electric Power Research Institute, Palo Alto, CA

Ford Motor Company, Dearborn, MI

General Electric Company, Fairfield, CT

Gillette Company, Gaithersburg, MD

Goodyear Tire and Rubber Company, Akron, OH

Hercules Inc., Wilmington, DE

Hoechst Celanese Corporation, Somerville, NJ

Morton Thiokol Inc., Chicago, IL

PSE&G, Newark, NJ

Shell Oil Company, Houston, TX

Texaco Inc., Beacon, NY

Union Carbide Corporation, Danbury, CT

U.S. Environmental Protection Agency, Washington, DC

Table 6
Environmental Fate Methodologies Reviewed

AERIS Model

GEOTOX Multimedia Compartment Model

Leaking Underground Fuel Tank (LUFT) Manual

MYGRT: An IBM Personal Computer Code for Simulating Solute Migration in Groundwater

Personal Computer-Graphical Exposure Modeling System (PCGEMS)

PCB On-site Spill Model (POSSM)

Preliminary Pollutant Limit Value (PPLV) Approach

Pesticide Root Zone Model (PRZM)

Risk Assistant/Fate and Transport (RAFT) Manual

Seasonal Soil Compartment Model (SESOIL)

Superfund Human Health Evaluation Manual

Table 7
Health Assessment Methodologies Reviewed

AERIS Model

California Site Mitigation Decision Tree Manual

GEOTOX Multimedia Compartment Model

Hawley's Approach to Assessment of Health Risk From Exposure to Contaminated Soils

Leaking Underground Fuel Tank (LUFT) Manual

Massachusetts Contingency Plan (MCP)

New Jersey's Soil Cleanup Criteria

Preliminary Pollutant Limit Value (PPLV) Approach

Risk Assistant/Fate and Transport (RAFT) Manual

Risk Assistant

Superfund Human Health Evaluation Manual

EXECUTIVE SUMMARY

Analysis and Environmental Fate Committee

CHESS's Analysis and Environmental Fate Committee reviewed several existing environmental fate models which can deal with petroleum hydrocarbons. The overall purpose of the review was to support the development of a consensus risk assessment methodology that could be recommended to risk managers in their development of options for addressing soil contamination. The objective of this endeavor was to identify those models that deserve further consideration for inclusion in a decision framework for soil cleanup levels.

The selection process which resulted in the committee's recommendation involved initial review of all models and supportive material by each committee member. Their comments were consolidated, summarized, and transformed into recommendations by the committee chairs into a chairman's report. This report was then submitted to committee members for concurrence.

This Committee's initial task was to review and evaluate eleven existing environmental fate models which can potentially address petroleum hydrocarbons in soil (Table 3).

These models were evaluated on their basis in science, applicability and site specificity, ability to address multiple environmental media, input data requirements, strengths, and weaknesses.

The usefulness of these models is dependent upon their ability to quantify the behavior of petroleum and its products as they exist in several forms. These forms include:

- Residual petroleum or residual saturation: the mass of petroleum retained by soil particles via capillary forces as a mass of petroleum migrates beyond a unit mass of soil.
- Free-floating petroleum or the pancake layer: the migrating mass of petroleum or petroleum product that has spread laterally over the water table.

- Dissolved petroleum or dissolved product: organic chemicals released from a bulk quantity and are now dissolved in soil water or groundwater.
- Vapor phase petroleum and petroleum products: high vapor pressure organic chemicals existing in soil air that are released from residual saturation, free- floating petroleum, or dissolved product.

Seven conclusions were derived from the review. First, none of the reviewed models addressed the depth of penetration of a bulk release of petroleum or petroleum product onto soil.

Second, none of the reviewed models addressed the formation of residual saturation due to a bulk release of petroleum or petroleum product onto soil.

Third, none of the reviewed models addressed the extent of the spread of free product in a soil-groundwater system.

Fourth, none of the reviewed models addressed the migration rate of free product in a soil-groundwater system.

Fifth, none of the reviewed models addressed the migration rate of vapor phase petroleum and petroleum products in unsaturated zone soil.

Sixth, none of the reviewed models adequately addressed the migration of dissolved organic chemicals in a soil- groundwater system.

In summary, two models were recommended for further consideration as potential tools to aid the risk assessment process: SESOIL (and PCGEMS/SESOIL) and POSSM. Both models address the concentration of an organic chemical dissolved in soil water migrating through unsaturated zone soil.

SESOIL is well recognized and accepted by that segment of the scientific community which utilizes soil-chemical fate models. SESOIL has been extensively validated and shown to work under a number of scenarios; it is utilized primarily in simulations requiring time periods greater than a month. SESOIL is compartmental, allowing for a significant amount of user tailoring of the model to a specific data set or site conditions.

SESOIL can accommodate the physical, chemical, and biological reactions of the chemical under scrutiny in soil systems. These reactions include:

- organic chemical reactions in soil (i.e., hydrolysis, general substitution, elimination, oxidation, reduction, soil-catalyzed reactions)
- organic chemical biodegradation in soil
- organic chemical adsorption

PCGEMS is a graphic exposure modeling system, capable of being run on an IBM-PC compatible computer, that incorporates SESOIL. PCGEMS houses a variety of programs that allow the user to estimate a number of environmental and chemical properties of organic chemicals to be used in SESOIL. PCGEMS utilizes SESOIL as the primary environmental model to characterize chemical movement through the soil system. Therefore, the strengths and weaknesses of PCGEMS/SESOIL as a tool rests primarily with the strengths and weaknesses of SESOIL.

POSSM is a contaminant transport model developed to predict environmental concentrations associated with a chemical spill. The model predicts daily changes in chemical concentrations on a spill site's soil and vegetation as well as losses of chemicals due to volatilization, surface runoff/soil erosion, and leaching to groundwater. The model was orginally developed for PCB spills.

POSSM is a one-dimensional, compartmental, dynamic, transport and fate model. It is one-dimensional because water and chemical movement in the soil are assumed to occur only in the vertical direction. It is compartmental because the soil column can be divided into a number of layers, each possessing different soil properties. The model is dynamic because a daily time step is utilized to calculate changes in hydrologic conditions and chemical concentrations.

The model mathematically considers important environmental processes including percolation, infiltration, runoff, evapotranspiration, and volatilization. It also considers degradation; however, one or more degradation processes must be expressed as a single first order degradation rate constant.

Environment and Health Committee

The Environment and Health Committee's purpose is to map a course for developing a toxicology-based approach to assessing the risks of petroleum contaminated soils to human health. The process must be solidly grounded in science, with each assumption presented, documented, and properly evaluated with each equation understood. At the same time, this process must be easily used by regulators and easy to modify on the basis of alternative site-specific assumptions by parties responsible for cleanup of the contaminated sites. It must neither be excessively simplistic ("one size fits all") nor require expensive customizing ("designer quality"); rather it should be a reasonable ("off-the-rack") methodology. It would be subject to timely modification as better information becomes available.

Of the extremely large number of constituents in the petroleum-based fuels of interest, only a limited number are of significant public health concern. These would be selected carefully, and their toxicity indices rigorously defined. The toxicological evaluations would be combined with site-related exposure assessments to produce the required risk assessments.

Since the beginning of 1990, the CHESS Environment and Health Committee has been examining documentation of a variety of methodologies that might be used. Expectedly, none was designed to solve the particular risk assessment needs posed here, yet all had certain features worthy of incorporation. To some extent, their faults or virtues are irrelevant to the present requirements; to a considerable degree, the faults are easily corrected. What follows are summaries of the important attributes and shortcomings of eleven "models" reviewed by the Environment and Health Committee.

AERIS

The "AERIS (An Aid for Evaluating the Redevelopment of Industrial Sites)" model, prepared for Environment Canada's Decommissioning Steering Committee by SENES Consultants, is strongly science-based. It employs a user-friendly interactive computer program to examine on-site human health risks of relatively old contamination — but not recent spills. Because of its on-site focus, it omits some environmental compartments, exposure pathways, or scenarios. Default values are Canada-orientated. AERIS is well documented, well written, has excellent graphics, and cites some very pertinent recent literature. It allows the user to either calculate risks from a site or to develop cleanup levels (thus going backwards and forwards). The emphasis is essentially on equilibrium conditions and lifetime exposures. AERIS lacks contaminant transport features and would not easily deal with complex mixtures. It would probably be hard to reprogram to fit within a comprehensive decision framework.

RAFT

The "Risk Assessment/Fate and Transport" modeling system, a product of the Pennsylvania Department of Environmental Resources, is based on sound scientific foundations. It seems to be broadly applicable — to most organics and some (but not all) inorganics, on-and off-site — but is not designed for complex mixtures or petroleum fuel products. RAFT is site-specific and the computer program adequately covers all media and most pathways (including those involving transport). Risks can be combined, but an integrated approach to risk calculation is lacking. The program is flexible, but not particularly user-friendly. The assumptions tend to be too conservative. The draft user's manual is hard to follow and incomplete, missing an entire chapter on input variable values; pathways and equations are not adequately described.

PPLV (PHAS)

The "Preliminary Pollutant Limit Value" (PPLV) concept, developed at the U.S. Army Biomedical Research and Development Laboratory, was computerized and documented. The computerized version is called "PHAS (Pollution Hazard Assessment System)". The PPLV approach is considered thorough, reasonable, and well based in science. It is pragmatic and has broad applicability, covering a large variety of pathways, but focuses on older contamination, where equilibrium conditions tend to prevail, and on the low concentrations that would remain following an effective cleanup. Thus, as presently formulated, it is not particularly applicable to petroleum fuel mixtures, especially not to spills. However, it is flexible enough to be reprogrammed to accommodate needed changes, including fate and transport aspects. The

PPLV approach is site-specific and addresses the various media and compartments. The PHAS program is flexible but needs to be user-tested; instructions are not well written. It would not be very useful for emergency response. A number of correctable deficiencies have been identified by the reviewers in its somewhat simplistic treatment of toxicology. Moreover, some of the default values (exposure factor assumptions) appear to be outdated, and newer sources of toxicological information, such as IRIS and the ATSDR's toxicological profiles, are not mentioned.

Risk Assistant

"Risk Assistant" is a computer program prepared by the Hampshire Research Institute for the U.S. EPA and the New Jersey Department of Environmental Protection. Its function appears to be the retrieval of information on properties of environmental contaminants and regulatory guidelines/standards, and use of these in baseline risk assessment. It is not petroleum fuel-directed. The program has a large data base to draw on, so that only contaminant concentrations need be provided. It seems to be user-friendly. The program is more chemical- than site-specific, and site-related parameters cannot be added. Risk Assistant does not address petroleum fuel contamination, or indeed complex mixtures generally. Transport is not considered. The manual that was provided is so lacking in information that Risk Assistant methodology could not be properly evaluated.

Hawley's Contaminated Soil Assessment

J. K. Hawley's "Assessment of Health Risk from Exposure to Contaminated Soil," *Risk Anal.*, 5:289–302 (1985) narrowly targets direct exposures to carcinogens in soil and house dust of residential areas, making no pretense to consider all probable exposure pathways. It may be viewed as a useful component of risk assessment methodologies of greater complexity and scope, though some of the rather conservative assumptions would certainly have to be modified, particularly in the light of recent research. Hawley's model is not very useful for risk assessment of petroleum fuel contamination because it contains no fate and transport component. It has not been computerized.

California Decision Tree

"California Site Mitigation Decision Tree Manual" was prepared by the California Department of Health Services. The manual has a strong basis in science with good referenced documentation. Sampling and analysis procedures are included. It covers a variety of sites, provides no scenarios, and is very California oriented. It is probably appropriate for petroleum fuel-contaminated sites, but would be cumbersome for field personnel or regulators to apply. Numerous pathways are missing and the document lacks detail. It does not lead to cleanup levels. It is not computerized.

GEOTOX

"GEOTOX—A Multimedia Compartment Model" by T. E. McKone et al., originated at the Lawrence Livermore National Laboratory. It is intended as a computerized screening tool for toxic environmental contaminants, and was designed to be a multicompartmental, partially open, time-varying chemical concentration system using landscape (not site-specific) data and the physicochemical properties of the chemicals of concern. Both because it is a global model and because of the continous-input aspect of its design, it would appear basically unsuited to relatively small-area petroleum fuel- contamination sites where the addition of more and more of the contaminants is not occurring. Its strong points are its ease of use, speed, thoroughness, and multi-media treatment. Its fate/transport equations also allow for chemical transformations, can follow a reaction product from medium to medium, and possess the capability for sensitivity analysis. There is some question as to its user friendliness. It appears, moreover, that many of the assumptions are embedded in the model and are not easy to reprogram. Incorporating missing pathways might be difficult. GEOTOX is not very applicable to petroleum fuel mixtures.

LUFT

The "Leaking Underground Fuel Tank Field Manual" was prepared by the California Water Resources Control Board. It is a "by-the-numbers," state-prescribed decision tool for cleanup of soil contaminated by diesel fuel and gasoline, and is based on the highly regarded SESOIL and AT123D models. As input, only analyses and limited geohydrology and soil characterization are required. The focus is on site-specific groundwater protection, and that purpose seems to be fulfilled. There is no computer model. The approach is consistent, practical, easy to follow, and acceptable for use in emergencies. Furthermore, the approach is narrow, in the sense that it considers no indirect pathways, and is in general extremely simplistic.

New Jersey's Soil Cleanup Criteria

"Soil Cleanup Criteria for Selected Petroleum Products", by S. K. Stokman and R. Dime, *Risk Assessment*, 342–345 (1986), is an evaluation and post-facto justification of a 100-ppm soil cleanup criterion for petroleum product-contaminated soils. This seems to be the only approach dealing with the health risks of petroleum products in soil. Based on literature data, it is primarily concerned with the exposure of children to surface soil contaminated by a variety of petroleum products. The only user input required is the analytical value for total petroleum hydrocarbons; the only chemicals of concern are carcinogenic polynuclear aromatic hydrocarbons (PAHs) and benzene. It is a quick-and-easy criterion, only a rule of thumb, and ignores both the concentration of PAHs and the loss of benzene that occurs when light fractions gradually volatilize away. Because of its conservative assumptions, it tends to overestimate risk. This paper contributes to our stock of ideas, but is in no way a risk assessment model. It is not computerized.

Massachusetts Contingency Plan

The "Draft Interim Guidance for Disposal Site Risk Characterization" was authored by the Massachusetts Department of Environmental Protection. It is consistent with Superfund guidance, employing standard scientific approaches. It emphasizes chemical analyses rather than physicochemical equilibrium models. It is intended to apply to all situations, but lacks the flexibility to be truly site-specific; it assumes that all sites are geologically and meteorologically equivalent. Since this methodology is not computerized, the user provides all input values. The procedures are straightforward, being based mostly on contamination levels. The references in Appendix B of the document are useful. The guidance in this document depends greatly on referenced State of Massachusetts publications. Only a few pathways are presented. No help is given for establishing cleanup levels. Overall, this guidance document provides a very simplistic approach to endangerment assessment.

RAGS-I-A

"Risk Assessment Guidance for Superfund, Vol.I — Human Health Evaluation Manual (Part A) — Interim Final" was produced over a number of years by the U.S. EPA. It effectively addresses a great many pertinent scientific issues, going into considerable detail. Unfortunately, it is not well supplied with source references. It is a marvelous resource — especially for on-site contamination — and is applicable to a broad variety of problems on a site-specific basis. RAGS-I-A is understandable, has good graphics, and gives insightful guidance; it is the de facto standard by which other methodologies are judged. Although a good reference manual and an excellent primer on environmental chemical risk assessment, RAGS-I-A would be difficult for the novice to use as a blueprint for conducting baseline risk assessments or defining cleanup goals. RAGS-I-A is cumbersome and unsuited to use for large, but especially for small petroleum contamination; it does not address mixtures. Inability to handle propagated uncertainity can lead to excessively conservative risk estimates, the more so because the assumptions used for exposure factors are overly conservative. The frequent referral to other documents, particularly with regard to exposure estimation, makes RAGS- I-A far less than the one-source guide many users would like to see.

Five of the 11 risk assessment methodologies reviewed above are associated with computer programs, namely AERIS, RAFT, PPLV (PHAS), Risk Assistant, and GEOTOX. Consideration was given to the utility of these programs for the matrix calculations, with the understanding that the programs might need considerable modification. The reviews indicated that GEOTOX would probably be too inflexible to modify, e.g., to change it from continuous to one-time input, or to use site- specific rather than landscape parameters. The Committee's knowledge of Risk Assistant was too vague, and to the degree it knew, the program was too unsuitable to favor further consideration. RAFT had been characterized as not particularly user-friendly, and the model and its assumptions had neither been well nor completely described. This left AERIS, a very user-friendly program, but perhaps not sufficiently modifiable, and PHAS. AERIS is flexible in allowing calculations of existing risks and risk-based cleanup levels. PHAS, on the other hand, may be a more adaptable system, as well as interactive, but has not been widely tested. CHESS has decided not to consider AERIS since its developers will not allow outside parties access to its computer codes or development of its algorithms in enough detail necessary to make a comprehensive evaluation.

Thus, it is the Environment and Health Committee's recommendation to select PHAS. PHAS will be modified under CHESS supervision so as to incorporate the views of the Expert Committees. With a suitable program in hand, development of equations and instructions to fill a decision matrix could proceed in stages, with frequent referral to CHESS for approval of assumptions and scenario/pathway selection.

CHESS will provide a reasonable number of sets of simple, but scientifically derived instructions, each set corresponding to the answers to a group of situation- and fuel-related questions.

Aid for the Evaluation of the Redevelopment of Industrial Sites[1-4]

Developer

SENES Consultants
Prepared for Decommissioning Steering Committee
Environment Canada

Available from
AERIS Software, Inc.
52 West Beaver Creek Road
Unit Number 4
Richmond Hill, Ontario

Phone
416/882-9106

Facsimile (FAX)
416/764-9386

Description

The AERIS model consists of four basic elements: a preprocessor, component modules, a postprocessor, and the supporting data base.

The AERIS system is designed to estimate an individual's inhalation and ingestion exposures to contaminants released into the soil of a particular site. Exposures are calculated using transport models and input data involving properties of the site environment and the contaminants of interest. The movement of a chemical from one environmental compartment to another is based on mass transfer coefficients in this model.

Environment Canada's AERIS model evaluates soil contaminated by past industrial activities, as opposed to active facilities or recent soil contaminant spills, and attempts to develop soil cleanup criteria in Canada based on these evaluations. The model creates scenarios involving a limited number of soil contaminants (e.g., benzene, phenanthrene, lead). Limited site-specific data can be utilized in place of default values to increase site-specificity. Human health concerns which are the driving factors in establishing methodologies are limited to on-site receptors. Toxicity factors and scenarios are utilized to develop acceptable concentrations of contaminants in soil. AERIS is limited in the number of contaminants evaluated and site-specificity; however, it is flexible by design and one can utilize more appropriate default values on specific contaminants if available.

The software is programmed using an expert system approach. A variety of fate models are used, mostly theoretical but claimed to have been validated. These models start with soil contamination levels and predict the concentrations in the following compartments: outdoor air, dust, indoor air, produce, and groundwater. The fate and transport models appear to be commonly used ones with the possible exception of plant uptake. These models are acknowledged as being the least well understood and presumably one of the most uncertain portions of the program. The produce levels are predicted from a combination of uptake via the roots and foliar deposition. Plant uptake does not account for losses that may occur during food preparation such as washing and peeling and does not account for adsorption of vapors. Exposure and risk calculations are made via standard approaches.

The computer model "uses various pieces of information needed to characterize a site redevelopment, estimates the potential doses that could result and determines acceptable concentrations of chemicals in soil which, in turn, can be used to derive cleanup standards" (stated objects). The model consists of databases containing information on chemical, receptor and generalizable site characteristics, computational algorithms to predict movement of chemicals from source to receptor and estimate doses, and an expert system based set of rules which control the flow of input data, computation, and output. The methods used to assess exposure and risk were developed in a series of workshops sponsored by the Decommissioning Steering Committee and represent a consensus of scientific opinion.

The scientific validity of the data bases supplied for the user of the model ranges from well-known physical and chemical parameters, to less generally accepted chemical toxicity values, to educated guesses regarding assumed and/or default input parameters. The computational methods are generally well specified in the text of the report or cited in references which include the open scientific literature, government reports and documents (including both Canadian and U.S.) as well as consultant's reports to the specific agencies. The method is generally similar to the approach recommended by the EPA for Superfund risk assessment.

The workshop, from which this report was developed, made as its highest priority the development of a method for establishing cleanup criteria. The workshop attendees recommended that "any method or approach used to develop cleanup criteria must include the following four requirements:

- site-specific parameters must be incorporated into the method
- quantitative consideration must be given to the interrelationship between air, soil, ground and surface water
- any method must not be specific to one type of contaminant (e.g., heavy metals, organics, indicator parameters or surrogates and
- provisions for examining various levels of protection (risk) for the environment and for human health must be incorporated into the model."

Scenarios are formulated, involving individual pathways, which are mathematically modeled. Compound-specific values are provided from data bases or estimated, largely from regressions based on similar compounds' properties. Site-specific data are required as input. The overall process is quantitative risk assessment.

Fate Evaluation

Basis In Science

There are four modules of interest: correlation, air, unsaturated zone, and saturated zone. The correlation module is used to predict air transfer coefficients. The predictions are subsequently used in the air module which calculates the flux of a chemical from the soil into the air and the basements of buildings.

The unsaturated zone module predicts concentrations in soil-water and in soil-air.

The saturated zone module predicts concentrations in groundwater. The rate which the chemical is transported is mathematically corrected for soil properties, chemical properties, and environmental conditions by utilizing mass transfer coefficients and partitioning coefficients.

Applicability

AERIS is designed to address chemicals in the unsaturated zone and the saturated zone and utilizes EPA's RITZ/VIP fate model. The model is not limited to application to any specific soil type or dissolved organic chemical.

The Vadose Zone Interactive Processes (VIP), an expanded version of RITZ, simulates the processes of volatilization, degradation, adsorption/desorption, advection, and dispersion. Four phases are considered with partitioning between them. The output consists of concentration profiles as functions of time and depth. Apparently, the water saturation is assumed constant throughout the zone and the water velocity is calculated using the recharge rate, porosity, and conductivity.

AERIS can be applied to a wide variety of sites, and it does not require substantial site-specific data for calibration.

Multimedia Relationships

AERIS is structured for integrated simulation of the unsaturated zone, saturated zone, and air compartments. This model does not consider the soil-groundwater-surface water- air compartments.

Input Data Requirements

Use of the model requires many input parameters. The reader is directed to the references at the end of the book for further information.

Strengths

It appears that AERIS does not require a user who is relatively "sophisticated." However, it may require input from databases containing substantial quantities of data.

The model does not require the use of a computer possessing capabilities greater than an IBM-PC compatible.

The model should be applicable on a wide variety of sites because it does not require substantial site-specific data for calibration.

Weaknesses

AERIS is not a well-recognized or well-utilized model that is readily accepted by that segment of the scientific community which utilizes soil-chemical fate models.

No data and information have been presented to show that the model has been extensively validated and shown to work under a number of scenarios.

The majority of the references cited in the supporting texts contain data, information, and approaches that are not current state-of-the-science.

Equations utilized to mimic chemical migration and transformations are not current state-of-the-science.

The model cannot accommodate on an individual basis the physical, chemical, and biological reactions of the chemical under scrutiny in soil systems.

The model depends heavily upon mass transfer coefficients, which are generally not measured for most chemicals of environmental concern.

The model addresses dissolved organic chemical movement in the unsaturated and saturated zones. Regarding free product, however, the model does not address:

- depth of penetration of bulk hydrocarbons
- spread of free product
- migration rate of free product
- effect(s) of large concentrations of other organics on adsorption and mobility
- emission rate for a pure bulk hydrocarbon on a soil surface

Comments

The model appears to be technically valid for the conditions for which it was developed. However, the model does not appear directly applicable to simulating hydrocarbon transport in the unsaturated zone from a surficial spill or leak.

The model contains a number of processes which are not directly related to hydrocarbon transport such as plant up-take, active treatment zone, etc. While the user could possibly ignore the nonhydrocarbon related factors with potentially little loss of accuracy, the resultant model would be a hybrid and potentially viewed as less credible than the original model.

Health Evaluation

Basis in Science

The approach developed by the authors of this model involves the determination of site-specific cleanup levels for soil by considering on-site exposure. Four land uses are considered as well as five exposure pathways. The exposure pathways include ingestion and inhalation of soil, ingestion of produce, inhalation of vapors, indoor ingestion and inhalation of particulates, and ingestion of groundwater. The model has input parameters for two organic compounds, including benzene and a phenanthrene, and for three inorganic compounds (i.e., zinc, lead, and selenium). The user can also input information on physical characteristics of different soil types, such as pH, as well as various meteorological conditions. Using different exposure algorithms and toxicity factors, such as the ADI or cancer risk factor, an acceptable concentration for a contaminant in soil is calculated for different exposure scenarios. Multiple exposure routes are considered in the calculation of this value.

Applicability

The AERIS model is designed for applicability to various soil conditions at a variety of industrial sites. The number and type of constituents discussed is limited. However, the model is designed such that it can be expanded (as planned by the developers) to include other soil contaminants that are constituents of petroleum products (e.g., toluene, xylene). The user can also input physico-chemical data for other compounds.

The model is applicable to situations in which equilibrium has been achieved; it is not useful for immediate spill response. Further, it seems to be most applicable to situations involving one or a few well-characterized chemicals.It is not clear how one would sum exposures or risks in order to model a complex mixture. The demonstration version of AERIS can handle both metals and organic compounds. Analyses are carried out compound-by-compound and mixtures are not addressed. The basic site simulated in AERIS consists of a surficial soil contaminated with a toxic substance. This contaminated zone serves as a constant input to a deeper soil zone, which in turn serves as a source to the saturated (ground water zone). Steady-state inputs are assumed and there is no chemical transformation due to biotic or abiotic processes.

AERIS has a specific application to contaminated industrial sites in which it is assumed all the contamination is contained within the physical boundaries of the model, the receptors are assumed to be present on the site, and the site is in a physico-chemical equilibrium state. It is not intended for use involving recent chemical spills, nor to calculate exposure to off-site workers or residents. It does consider the future land use of the site to be either residential, industrial, agricultural, or recreational (park land), again with the assumption that all exposures will be to workers or residents on-site.

The current version of AERIS contains physico-chemical data on approximately 40 compounds, including common volatile organics, pesticides, PCB, TCDD, but only three metals (lead, zinc, and selenium). The data is not complete for all compounds, e.g., half-lives in soils must be manually input. Adding additional chemicals requires the LEVEL 5 expert system software which controls the flow of information between databases and computational algorithms. Additional metals also require new algorithms to be developed. Metals such as arsenic with complex chemical-soil interactions would require a substantial level of effort.

Emphasis is on the human health effects, under Canadian

ambient conditions, of individual chemicals. Although the orientation is towards determination of cleanup levels, rather than towards endangerment assessment, the claimed capability of forward and backward calculation would permit use for endangerment assessment. The methodology focuses on on-site contamination that is not of recent origin. It is not petroleum product- or mixture-oriented.

The basic assumption of the AERIS model is that site-specific data is available to guide risk management decisions on a site-specific basis. In addition, the user can choose not to use default values and incorporate alternative site-specific factors for a particular chemical. Examination of the input parameters shows the obligatory dependence on this data. Although the expert system interface can supply default/recommended values for many parameters, the model cannot be used without the following types of site-specific data: (1) local climate, (2) soil type, texture, and properties, (3) potential receptors, (4) groundwater usage, (5) possible exposure pathways, and (6) future land use.

The program's requirement of numerous input values for site characteristics indicates that it could be applied to a wide variety of sites. The AERIS model incorporates a substantial amount of geological and meteorological information. At present, it has available considerable information on various locations in Canada. Input requirements include meteorological data (data sets are available for the user), soil characteristics, groundwater characteristics, and chemical properties. AERIS includes resident databases of various parameters, which the user can edit to obtain relevant site-specific values. The database is limited however, to 4 organics, 3 inorganics, meteorology for 6 cities, 9 soil types, and 14 geologic formations. This database is used to create defaults, so the site applicability is limited when the user relies on the default values and chooses the most similar Canadian site already in the database.

As with most models, AERIS involves a series of simulations and exposure scenarios as opposed to evaluations based on real site data; thus, site specificity is compromised and based more on theory. The model is flexible and can address various specific soil types and chemical-specific information involving established default values. Alternate default values can be utilized in the model if they are more appropriate than the values cited for the contaminants in the AERIS scenarios.

Multimedia Relationships

AERIS addresses the movement of chemicals between surficial soil and adjoining media (e.g., deeper soil, groundwater, air, and household (basement) air). Off-site transport to surface waters is not explicitly considered, as the AERIS formulation deals primarily with site redevelopment.

The model considers the following exposure pathways: direct ingestion of local soil and indoor air, ingestion of groundwater from an adjacent well, and ingestion of produce grown in local soil. Potentially important pathways not considered (reflecting the industrial nature of the site) include dermal absorption from soil contact, ingestion of fish and/or locally grown meat (because AERIS envisions primarily urban uses), and inhalation of vapors while showering.

It is assumed that there are no surface water bodies that might be used for drinking or recreation, and as a consequence, water-to-air transfer is assumed not to occur. The model assumes that contamination starts at the surface. It does estimate transfer from soil to groundwater, air above the ground, and air below the surface of the ground (i.e., into a basement). As a consequence, water to air transfer is assumed not to occur. It does consider transfer from groundwater to humans directly.

Consideration of these pathways requires many different computations of movement of chemicals from one media to another. The model is similar to compartmental type models in that it produces separate calculations for saturated soil, unsaturated soil, outdoor air, and indoor air. The environmental fate and transport models used are generally simple, but are well documented both in the AERIS report and in open scientific literature. No attempt is made to validate fate and transport model predictions of exposure point concentrations. The user is presented with the option of supplying a measured or estimated concentration in the air, drinking water, and/or food consumption pathways.

AERIS considers all environmental pathways pertaining to soil, except for the migration of soil contaminants as runoff (leachate, ponding). In addition, the model does not address off-site contamination or exposure of off-site receptors to contaminants that have migrated off-site. In this regard, the model does not address off-site groundwater contamination from the migration of contaminants from soil to groundwater and subsequently, it does not address the ingestion of contaminated groundwater by private well users downgradient from the site. Off-site contamination needs to be addressed in the future in order to adequately address public health concerns.

Input Data Requirements

As noted above, inputs required to run AERIS are derived from two sources: user-defined inputs and inputs from AERIS databases as well as default values.

Compared to many of the refined environmental fate and transport models available in the literature (e.g., SESOIL), the data requirements for AERIS would be considered moderate. One of the most attractive features of AERIS is an expert system interface which can draw input from several built-in databases to supply default and/or recommended values for many (but not all) of the required input values. The following categories of input are present in the model: (1) chemical properties, (2) toxicological information, (3) environmental criteria and adverse effects information, (4) receptor information, (5) receptor activity patterns, (6) meteorological information, (7) unsaturated soils data, (8) saturated soils data, and (9) crop production/consumption values. The user has the option to edit and change any of these values before the computational sequence is initiated.

The databases supplied in the prototype version of AERIS (in DBase III format) are specific to the Canadian environment (e.g., number of days spent outdoors in winter, soil types, etc.). For AERIS to be used outside of this physical universe, many of the necessary input parameters will need to be supplied by the user without the assistance of the expert system knowledge base. Two important classes of parameters which are specific to this universe cannot be easily altered: residential houses with a fixed dimension basement that drives the indoor air pathway calculation, and the parameters associated with production, contamination and consumption of crops grown on-site.

At a minimum the AERIS methodology requires physico-chemical data including CAS number, molecular weight, vapor pressure, log-octanol/water partition coefficient, solubility in water, liquid diffusivity (although a default value of 1.8×10^{-6} is provided), air diffusivity, and half life in soil. The receptor information is provided as a series of default values. Site parameters including depth of soil contamination and length of site in the direction of groundwater movement must be provided by the user. Information would also include demographic information and anticipated land uses for choice of scenarios. Values of other parameters are default values fixed by the scenario selected.

Strengths

The AERIS methodology has done a reasonable job of identifying potential routes of exposure, has realistic scenarios (at least within the context of its mission), and considers foliar deposition on plants in addition to uptake via roots. In general, it seems to be a very good method of estimating the potential future exposure from a site contaminated with one or a few well-characterized chemicals.

The AERIS model allows duration and time of exposure to be included in risk calculations of carcinogens. It attempts to deal with the troublesome questions of defining carcinogen exposure levels that present a health risk, and deciding how those exposure levels should be averaged over a lifetime. The model permits the user to allow for less than 100% bioavailability.

The most visible strength of AERIS is its ease of use, which stems from the use of a well-known expert system software, LEVEL 5. This system controls the flow of input, supplies many default and/or recommended parameter values, and performs computations, based on replies to questions or queries presented to the user. Additionally, the databases provided and the knowledge base used by the expert system are substantial and represent a considerable amount of scientific input and thought.

Supporting documentation is adequate to understand the nature and rationale of the input parameters and computations performed, but not adequate to allow users to maintain, modify or augment the model capabilities. The model is also well documented in the open literature.[4]

The AERIS model explicitly deals with important soil-based exposure pathways. The algorithms adopted for use in the model have been partially verified against scenarios reported in the open literature. The interactive nature of the system is attractive because it facilitates the analysis of alternative chemicals for various site-specific conditions. It can incorporate land use considerations in an expert system format, aids the estimation of unknown inputs, models diffusion of vapors into basements and subsequent exposures, and graphically displays output, such as pie charts on relative contributions of exposure pathways.

The AERIS model may serve as a good prototype for exposure assessment of contaminated soil and the basis for the development of appropriate guidelines for soil cleanup in the interest of human health. Several exposure routes are considered (except for dermal contact). One of the greatest strengths of the model is the inherent flexibility in allowing modifications in the default parameters, thus increasing site specificity.

Weaknesses

The single chemical approach limits utility for complex mixtures.

Physical chemistry values such as partition coefficients on a complex mixture, such as petroleum products, may be impossible to obtain. Partition coefficients of individual components will not reflect behavior of the mixture.

The system is designed to evaluate only those sites where contamination has reached equilibrium. This implies slowly changing contaminant concentrations and not active sources. Thus, this simulation approach would be inappropriate for examining hydrocarbon concentrations in surficial soils that are changing as a function of volatilization and/or biodegradation. The model, therefore, has limited utility for emergency response.

The failure to evaluate migration off-site may result in underestimating hazards to off-site receptors.

The failure to consider surface water or aquatic organisms may result in underestimating human exposure and/or environmental hazard.

The limited number of scenarios may not accurately reflect a situation in which a company may wish to clean up a contaminated site on property not under consideration for alternative uses.

The small number of chemicals in the current database requires the user to obtain considerable physico-chemical data on other materials to run the model.

The AERIS model does not consider dermal exposures to contaminants in surface soil.

The AERIS program should be altered to handle distributions of values for some of the constants, such as body weight, age, and amount of air breathed.

In addition, it does not appear that the module allows one to reduce the exposure period as an adult to a less than lifetime exposure. When one considers that the median length of residence in the United States is 9 years and the 90th percentile value is 30 years, consideration of time can be very important.

AERIS does not include values for amounts of meat or milk ingested. It also does not appear to be as flexible or able to handle as extensive a set of exposure scenarios as the PPLV approach.

Data on which virtually safe doses (VSDs) are based and derived is not adequately documented. Produce (plant uptake) models are not well validated.

The assumption of strict additivity when calculating hazards from multiple materials may result in unreasonably low levels of exposure for individual components.

The AERIS model offers overly conservative default values. For instance, 250 mg/d of soil is estimated as the amount of soil that children ingest. However, this value is higher than the EPA's current value of 200 mg/d for children ages one through six and is also higher than the value that Calabrese et al.[4a] found in their soil ingestion studies (9 to 40 mg/d), does not adequately address inhalation of respirable soil particles,[5] and the number of constituents evaluated is too small and should incorporate more constituents found in petroleum products.[6]

Comments

It is not clear to what extent the assumptions made in the document are supported in the literature, and the following specific assumptions are made regarding the model:

1. One assumption concerning ingestion of produce is that contaminant levels in plants will be proportional to those in groundwater.
2. The assumption concerning direct oral ingestion is that bioavailability is 100%. This may overestimate actual absorbed dose by an order of magnitude.
3. The assumption concerning inhalation of total suspended particulate (tsp) is that 100% of the tsp is respirable. A second assumption is that bioavailability is 100%.
4. The key assumption concerning dermal absorption is that this route is negligible, which may or may not be true. The problem is that dermal contact (which may represent the greatest hazard for high boiling point petroleum products) is ignored.

There is no discussion about the uncertainties inherent in default values. For many of the assumptions presented, no data or references are given or the reference is an obscure and most likely unreviewed report.

One of the chemicals used as an example is phenanthracene. An acceptable soil concentration given in the report for phenanthracene is 0.3 mg/kg. This is supposed to be associated with an air concentration of 0.1 mg/m^3. It is doubtful that an air level of 0.1 mg/m^3 is associated with a soil level of 0.3 mg/kg of soil.

Using the model, the number developed as a cleanup goal does not seem useful or realistic, particularly when the default assumptions are used. There is no reason to expend resources cleaning up a site to these low levels when the resulting exposure from levels that are several orders of magnitude higher would only represent a fraction of less than 1 to 10% of the overall exposure humans receive from other sources. The default assumptions given for human uptake are for the most part not realistic. It appears that little effort was made to base assumptions on available data. To make this model useful, extensive revision is necessary.

At this time, Environment Canada has not officially released AERIS to the public. Without access to the original code, databases, and expert system knowledge base, it is not possible to verify any of the calculations produced. Consequently, AERIS is not useful to other organizations, public or private, since there is no audit trail for calculations, decisions, or recommendations produced. In addition, the LEVEL 5 expert system software package is required for any maintenance or modification to the current model.

To be generally useful, AERIS must provide a method for other users to incorporate their own site-specific assumptions, fate and transport models, and regulatory standards database, at a much greater level of detail than that provided by simply changing some of the input parameters.

Overall, the report seemed easy to read and presented graphical information that would be very helpful to the risk assessor. In addition, the input data is relevant to soil contamination. However, prior to recommending its use for a state regulatory official, several modifications must be incorporated. The exposure routes should be broadened to consider off-site exposure and additional exposure routes such as ingestion of contaminated meat or milk products and dermal absorption. The default values for exposure parameters should be modified to reflect more median values. Finally, input data on chemicals such as xylene, toluene, or other PAHs should be incorporated to make the approach more generally usable for petroleum contaminated sites.

SESOIL

Seasonal Soil Compartment Model[5-7]

Developer

U.S. EPA Office of Toxic Substances

Available from

U.S. Environmental Protection Agency
Exposure Assessment Branch
Room E., Tower 322
401 M Street, SW
Washington, D.C. 20460

Phone

202/382-5588

Description

The Seasonal Soil Compartment Model (SESOIL) was designed for long term environmental, hydrologic, sediment, and pollutant fate simulations. The model uses a theoretical soil column (compartmental) approach, and up to four separate homogenous soil layers can be differentiated. The model incorporates the hydrologic cycle, the sediment washload cycle, and the pollutant fate cycle. The hydrologic cycle considers rainfall, surface runoff, infiltration, soil water content, evapotranspiration, and groundwater runoff. The sediment cycle relates to sediment washload as a result of rainstorms (i.e., soil erosion from surface runoff). The pollutant fate cycle includes convective transport, volatilization, adsorption/desorption, chemical degradation/decay, biological transformation, hydrolysis, photolysis, oxidation, and complexation with metals.

The model assumes that pollutant concentrations in all phases and in all compartments of the soil system are at equilibrium at all times. Equilibrium concentrations of the chemical species are found by applying the law of mass conservation over a series of monthly time steps.

The hydrologic cycle module provides the "driving force" needed to run the pollutant fate cycle. It is based on a "statistical dynamic formulation of vertical water budget at a land-atmospheric interface" method. The method is used to calculate the average seasonal infiltration, runoff, evaporation, and soil moisture content. The "seasonal" approach is useful because it requires relatively few input parameters. The data can be acquired from readily available National Oceanic and Atmospheric Administration (NOAA) annual and monthly weather summary reports.

The SESOIL user's manual contains descriptions of the theory and equations for all processes.

Fate Evaluation

Basis In Science

SESOIL is a dynamic soil compartmental model containing a hydrologic cycle and pollutant cycle compartments, designed for long-term environmental pollutant fate simulations. These compartments and their utilization are extremely well supported by published data and information found in the scientific literature.

The model is based on stochastic representation of the hydrologic cycle in which predicted water movement occurs at seasonal (monthly or yearly) intervals. The hydrologic cycle in SESOIL is based on a statistical dynamic formulation of the vertical vadose zone water budget at the soil-atmosphere interface. Physically-based dynamic and conservation equations express the infiltration, transpiration, percolation to groundwater, and capillary rise; independent input variables driving these equations include precipitation, potential evapotranspiration, soil properties, and water table elevations.

SESOIL uses a water or mass balance formulation to make explicit calculations for each soil layer for each time increment. Uncertainty of the simulation of the hydrologic cycle is incorporated into SESOIL via the probability density functions of the climatic variables.

Applicability

SESOIL is a useful screening model for determining the fate of a chemical in the unsaturated zone. The model is most successful for long-term studies since the hydrology is based on a water balance over a season rather than changes in soil moisture through time. The output calculates the chemical distribution in a soil column at the end of a month or year. The model is compartmental with the soil column having up to ten layers. Each layer is assumed to have uniform properties.

The model can be applied to generic environmental conditions for the purpose of evaluating the general behavior of a chemical. The model is not limited to application to any specific soil type or dissolved organic chemical.

The model is written in FORTRAN.

Although the model can be applied to a wide variety of sites, the model requires site-specific data for calibration.

SESOIL does not consider soil volume or area of spill. It assumes all of the chemical is incorporated to a certain depth. The model also does not consider initial concentration of the chemical.

The hydrologic cycle accounts for rainfall, infiltration, surface runoff, evapotranspiration, groundwater runoff, snow pack/melt, and interception.

The sediment cycle accounts for sediment resuspension resultingfrom wind and sediment washload resulting from precipitation.

SESOIL considers depth to groundwater table, ratio of the compound concentration to the maximum solubility, and thickness of up to ten layers.

Other site-specific requirements are located under the input data requirements. The input requirements are straightforward and the only elusive parameter is the disconnectedness index (i.e., variation in permeability with soil moisture derived by curve-fitting).

Multimedia Relationships

SESOIL is structured for the integrated simulation of three cycles: the hydrologic cycle, the sediment cycle, and the dissolved pollutant cycle. Therefore, this model considers interactions among the soil-groundwater-surface and water-air compartments.

SESOIL considers all phases to be in equilibrium. The model considers transport in the unsaturated soil (i.e., not in the saturated zone). The compounds of concern exist in the soil moisture, on the soil matrix, and in the soil air.

Input Data Requirements

Use of the model requires approximately 50 input variables, which is substantially fewer than required by many other models. Input variables can be classified into five categories: climate data, soil data, chemical data, initial chemical distribution in the soil, and washload data.

Specific Input Requirements:

1. Chemical:
   - adsorption coefficient based on organic carbon content
   - overall adsorption coefficient
   - degradation rate
   - solubility
   - Henry's Law Constant
   - number of moles of ligand per mole of compound complexed
   - stability constant of compound-ligand complex
   - valence
   - neutral hydrolysis constant
   - base hydrolysis constant
   - acid hydrolysis constant

- diffusion coefficient in air
- molecular weight

2. Soil characteristics:
   - soil bulk density
   - intrinsic permeability
   - disconnectedness index
   - porosity
   - organic carbon content
   - Freundlich exponent
   - cation exchange capacity
   - soil layer area
   - depth of groundwater
   - soil layer depths
   - pH
3. Climate:
   - temperature
   - cloud cover
   - relative humidity
   - albedo
   - mean precipitation
   - mean storm duration
   - number of storms
   - latitude
   - mean length of rain period
   - evapotranspiration rate

Precipitation is statistically computed based on historical rainfall data. Surface runoff, groundwater, infiltration, evapotranspiration, and drainage are based on the precipitation. Infiltration is described by the Philips equation which assumes that the Richard's equation can be applied to a semi-infinite porous medium.

SESOIL contains no internal checking of the input data.

Strengths

The model is well recognized and accepted by that segment of the scientific community which utilizes soil-chemical fate models.

The model has been extensively validated and shown to work under a number of scenarios. The model is utilized primarily in simulations requiring time periods greater than a month.

The model can accommodate the physical, chemical, and biological reactions of the chemical under scrutiny in soil systems. These reactions include:

- organic chemical reactions in soil (i.e., hydrolysis, general substitution, elimination, oxidation, reduction, soil-catalyzed reactions)
- organic chemical biodegradation in soil
- organic chemical adsorption

SESOIL also considers photolysis, nutrient cycles, and complexation chemistry.

SESOIL has been tested and some conclusions on the model indicate that the predictions made by SESOIL are better for long-term rather than short-term analyses. Arthur B. Little, Inc. compared predictions made by SESOIL with field data for metals and organics. Sensitivity analyses were made on the adsorption and volatilization for different soil types and in different climates. Battelle Laboratories conducted a study of SESOIL and other models for soil and groundwater systems.

SESOIL is supported by the Office of Toxic Substances of EPA. Corrections and improvements are continuously being made on the model by Oak Ridge Laboratory.

The hydrologic cycle of SESOIL has been found to be a good long-term predictor for groundwater and surface runoff, evapotranspiration, and infiltration.

Uncertainty analysis is introduced into the hydrologic cycle with probability density functions and produces probability distributions of water balance elements that yield long-term seasonal averages of the water balance (new version of SESOIL).

The model is compartmental, allowing a significant amount of user tailoring of the model to a specific data set or site conditions.

SESOIL is part of the PCGEMS package for exposure assessment studies. The package has large data sets which may be valuable.

Weaknesses

The model estimates the hydrologic cycle components from NOAA, USDA, and USGS databases. Therefore, the model requires a user who is relatively "sophisticated" and input from databases containing substantial quantities of data.

The model requires the use of a computer possessing capabilities greater than an IBM-PC compatible (note: A PC version was developed after this report was developed).

The model may not be applicable on a wide variety of sites because the model requires substantial site-specific data for calibration.

The model cannot be utilized on sites with very large vertical variations in soil properties. The model can accommodate up to four soil horizons. However, a single homogeneous soil column is utilized in the hydrologic cycle.

For input variables not readily available to the model user, very complex calculations will be required to obtain this data. These calculations and the decisions made to select the appropriate equations for use will require a most sophisticated user.

The model addresses dissolved organic chemical movement in the unsaturated zone. Regarding free product, however, the model does not address:

- depth of penetration of bulk hydrocarbons
- spread of free product
- migration rate of free product
- effect(s) of large concentrations of other organics on adsorption and mobility
- emission rate for a pure bulk hydrocarbon on a soil surface

The model considers only the migration of a single solute in an aqueous phase and does not consider the nonaqueous phase. Significant error can occur in a model that does not distinguish between the NAPL phase or the water phase as being the dominant transport carrier. POSSM does not consider this distinction.

Model testing in a laboratory setting did not find very good predictable results with SESOIL. The results did not improve when using laboratory derived input data. Other models were able to predict the effluent concentrations better than SESOIL.

According to some investigators, SESOIL was not able to work when there were low permeability layers in the soil column. Attempts to calibrate SESOIL were unsuccessful in this type of situation. In another site tested with SESOIL, the model predicted that 65.6% of the chemical would leach to groundwater when the field data showed that only 4 to 8% had leached at this particular site.

SESOIL has been "improved" since these studies were conducted, and the model has been tested in laboratory columns with six organic compounds and the predictions compared with three field tests. Better results were found with the improved SESOIL in that the predictions agreed more closely for some compounds. However, results for compounds with large adsorption coefficients did not show much improvement. Also, the compound with the lowest adsorption coefficient did not show much improvement.

For SESOIL to work optimally, the data needs to be calibrated specifically for the site. When literature data is used the predictions are not as accurate. This may be a major limitation for this model and the other models under review since good site-specific data may not be available and using literature or default values may produce predictions that are inaccurate by orders of magnitude. Extensive calibration is required to use this model on a site-specific basis.

SESOIL compartmentalizes the chemical concentration up to ten layers in the recent model, but the hydrologic cycle considers the soil column as only one compartment. Therefore, only one soil-water content porosity and core-disconnectedness parameters are used to describe the entire unsaturated zone. SESOIL has no spatial resolution when water content is being calculated. There is no resolution of the flow in each compartment. This could affect the chemical transport and distribution in the soil column since the retardation coefficients and volatilization fluxes of certain chemicals are water sensitive.

SESOIL also assumes the internal soil moisture at the beginning of each storm and interstorm period to be uniform at its long-term space-time average. This latter assumption may represent a considerable departure from reality, since the soil moisture profile in a later modeling period (i.e., one month) cannot be influenced by the soil moisture profile resulting from a preceding modeling period.

The sediment cycle of SESOIL does not consider chemical transport with the sediment. The sediment cycle has been tested less than the other cycles.

The layered approach of SESOIL has a weakness in that when a chemical enters a layer it is assumed to be instantaneously distributed uniformly throughout the entire layer. Therefore, the larger the layer, the more dilute the predicted chemical concentration.

The model ignores diffusive mobility of chemicals which may be important depending on the Henry's constant of the compound, and does not consider the volatilization enhancement when water evaporates. The study conducted by Hetrick et al.[7a] found that SESOIL underestimates the concentration near the surface and states that this may be a result of ignoring the upward movement of compounds. Hetrick et al. concludes that SESOIL would be a good screening level model.

SESOIL does not allow for below freezing temperatures.

The interaction between other chemicals present and the chemical of concern is not considered.

PCGEMS: Personal Computer Version of the Graphical Exposure Modeling System[8]; Seasonal Soil (SESOIL) Model

Developer

General Services Corporation/U.S. EPA

Available from

Annette Nold TS 798
U.S. EPA
401 M Street, SW
Washington, D.C. 20460

Description

PCGEMS is a graphic exposure modeling system, capable of being run on an IBM-PC compatible computer, that incorporates SESOIL, a dynamic soil compartmental model containing hydrologic cycle and pollutant cycle compartments.

PCGEMS houses a variety of programs that allow the user to obtain pertinent data to be utilized in models supporting exposure assessments. More specifically, its primary value is estimating a number of environmental and chemical properties of organic chemicals.

PCGEMS utilizes SESOIL as the primary environmental model to characterize chemical movement through the soil system. Therefore, the strengths and weaknesses of PCGEMS/SESOIL as a tool rests primarily with the strengths and weaknesses of SESOIL. The reader should consult the fate and transport evaluation in the SESOIL section.

California Site Mitigation Decision Tree Manual[9]

Developers

California Department of Health Services
Toxic Substances Control Division (May 1986)
Leu, D.J., Ph.D., M. Kiado, P. Hadley, S. Lau, J. Polisini, S. Reynolds, R. Sedman, S. Solarz, J. Tracey, and C. Woodhouse.

Available from	*Phone*
California Department of Health Services P.O. Box 942732 Attn: Katherine Clark (Decision Tree) Sacramento, CA 94232-7320	416/882-9106

Description

The California Site Mitigation Tree 1 consists of four major steps: (1) identification of toxic substances, (2) determination of Applied Action Levels (AALs), (3) environmental fate modeling, and (4) risk analysis.

The identification of toxic substances involves review of chemical information concerning substances at the site. The available data are evaluated for quality and adequacy. The chemicals of interest are classified as carcinogens or noncarcinogens.

AALs are defined as media-specific levels of a substance which, if exceeded, present a significant risk. The AALs are derived from No Observed Adverse Effect Levels (NOAELs) for noncarcinogens and from risk factors for carcinogens. Standard body weight, standard intake, pharmacokinetic, and uncertainty factors are included in the derivation of the AALs.

Mathematical models are utilized to determine the environmental fate of a chemical of interest. Soil adsorption, bioconcentration, and emission rates are evaluated in this step. Exposure levels determined from the environmental fate analysis are used in the risk analysis. The risk analysis involves comparison of exposure levels with AALs for specific chemicals and media. Multimedia exposures and additive toxic effects are taken into account. If exposure levels are found to exceed the AAL for any chemicals, a significant risk is deemed to exist.

Health Evaluation

Basis in Science

The purpose of this decision tree manual is to provide a framework for establishing site-specific mitigation criteria. In addition, the methodology cited in the decision tree manual will select the remedial option(s) which most effectively addresses environmental and public health concerns by considering various site-specific variables (e.g., demographics, local concerns). The primary deliverable is to establish AALs for contaminants in various environmental media (soil, water, air). The AALs are biological receptor-specific as opposed to taking into account site-specific parameters.

Applicability

The manual is designed to apply to any site, from small and simple to large and complex. It can be used to evaluate any type of contaminant.

The manual represents itself as being applicable to any remediation action in California. However, the features which would make it applicable to California pertain to the assessment of the extent of contamination and to the site-specific geologic and meteorological characteristics.

Despite a very brief chapter (Chapter 6) on risk assessment applied to aquatic biota, the focus is on human health effects baseline risk assessment of individual chemicals, summed as total hazard indices and carcinogenic risks. Scenarios are not developed. The methodology is not petroleum- or mixture-oriented.

Food chain pathways and inhalation exposure to volatiles from tapwater are not covered. Instructions are missing for the utilization of the many useful approaches that are presented.

The Decision Tree process was designed to be applied to a variety of sites (e.g., small simple sites and large complex sites). Mitigative actions are based on site-specific parameters and the AALs are based on biological receptor specificities.

The abundant presentations of site-specific exposure-related calculations are not adequately incorporated into the sketchy outline of risk determination procedures of Section 8.7.

Multimedia Relationship

The decision tree is applicable to multimedia environmental fate and transport between soil, water, and air.

The presentation is somewhat uneven. There are instances of extremely thorough treatment, which seem to cover most situations, but some there are also important omissions, e.g., the difference between sediment and soil, bioaccumulation vs. bioconcentration, and the behavior of semivolatiles on airborne particulates.

The manual discusses measurements of the extent of contamination in soil, air, and water so multimedia relationships are recognized. However, the manual does not contain methods to estimate partitioning into various environmental media. Rather, the manual seems to take the view that predictive methods are subject to such uncertainty that only measured values should be considered.

Input Data Requirements

Since there is no associated computer software, all data must come from external sources, except such values (e.g., constants) as are supplied with the numerous physico-chemical equations.

Site-specific parameters (e.g., demographic information) as well as exposure input data and chemical specific information are needed.

Strengths

The manual provides good discussion of fate (from an exposure standpoint) and risk issues. The manual discusses how to set AALs. This tier style has the advantage of lending itself to use by agencies. The decision tree process probably could be easily computerized. It could be useful for emergency response if modified.

The attention paid to details of the site assessment and to explaining the development of reference doses is helpful. Development of exposure level equations is quite thorough.

This procedure is designed to be used at any hazardous waste site and to allow the user to incorporate new techniques as they become available. Unique features include sections on quality assurance and quality control, and evaluation of organisms other than humans. Helpful flow charts clearly delineate overall processes. A strong section (Component II: Site Assessment) deals with soil/water sampling and quality assurance in the hydrogeologic context. Models for environmental transport are well developed and abundantly referenced. Thus, the manual provides for use of reliable data with good models. It is probably very appropriate for petroleum contaminated sites with minor modifications.

Weaknesses

The decision tree process is too complicated to be useful to most field people, and provides insufficient detail to conduct a complete assessment.

The "decision tree" approach appears to be inadequately performed. No unifying concept is presented to lead the user through the maze of topics of varying complexity. For example, it would be difficult for a novice to develop allowable exposure levels — even for a simple nonvolatile nondegrading compound — for beef grazing in a contaminated field or watered in a feedlot with contaminated groundwater.

Human exposure to volatiles in tapwater is not covered. The behavior of volatile or semivolatile compounds on airborne particulates is not addressed. There is no discussion of the penetration of homes by soil volatiles through cracks in basement floors and foundations. Outdoor vapor exposure models are not discussed and compared with each other. The models are not designed for petroleum product mixtures. There is no supporting computer software. Most significantly, the user is not led to cleanup levels, and only poorly to baseline risk estimation. More detailed site-specific parameters are needed.

The techniques described here were developed or selected to meet California's needs; some may contradict regulations in other jurisdictions. Because California's environmental regulations are fairly stringent, the level of effort presented in this manual may be have to be substantially modified for sites in other states.

Comments

The manual seems to have been written primarily for engineers. The most detailed sections refer to measurement and assessment of the extent of site contamination. The sections dealing with exposure assessments and human and environmental risk assessments are short and relatively inflexible. In general, the risk assessment methodology consists of selecting a critical value and comparing that level to the exposure estimates. Thus, the risk assessment process is reduced to a calculation based on "regulatory values" or predictive equations.

Leaking Underground fuel Tank Field Manual[10]

Developers

David J. Leu and Terry Brazell

Available from

State of California
State Water Resources Control Board
Paul R. Bonderson Building
901 P Street
Sacramento, CA 95801

Phone

Diane Edwards
916/324-0911

Description

The State of California Department of Health Services and State Water Resources Control Board established procedures for determining whether an underground storage fuel tank site poses a risk to human health. The field manual is a practical extension of the "California Site Mitigation Decision Tree" document produced by the California Department of Health Services, Toxic Substances Control Division.

The guidelines presented in the Leaking Underground Fuel Tank (LUFT) Field Manual apply only to soil and groundwater contamination, rather than surface water contamination or air pollution. In addition, the guidelines only deal with gasoline and diesel fuel products and do not consider practical guidance to field personnel responsible for dealing with all leaking fuel tank problems.

For situations in which gasoline contamination of soil may have occurred, the manual suggests analyzing for benzene, toluene, xylene, and ethylbenzene (BTX and E), and total petroleum hydrocarbons (TPH). For situations in which diesel fuel may have contaminated a site, only TPH is measured.

The manual cautions against measurement of ethylene dibromide (EDB) and organo-lead for the following reasons. EDB has been in such widespread use that its detection may not be due to gasoline. Concerning organo-lead, most laboratories have the ability only to analyze for total lead, and cannot distinguish between inorganic and organic lead. In addition, inorganic lead is native to California soil, which may cause to false positive readings unless background levels are known.

The reasons offered by the manual for choosing BTX and E as indicators of gasoline contamination include the following:

- they are readily adaptable to gas chromatograph detection.
- they pose a serious threat to human health (i.e., benzene as a carcinogen).
- They have the potential to move through the soil and contaminate groundwater.
- Their vapors can be highly flammable and explosive.

Because BTX and E are highly mobile and can migrate from the site, it is also important to analyze for TPH. TPH detection is reported as the total of all hydrocarbons in the samples. Tank sites are initially classified as one of the following:

- Category I: No Suspected Soil Contamination, i.e., sites in which tanks are being closed for reasons other than a leak.
- Category II: Suspected of Known Soil Contamination, i.e., sites where tanks or lines have failed precision tests, show discrepancies in monitoring records, or show visual evidence of leakage.
- Category III: Known Groundwater Contamination, i.e., sites where tanks or piping have shown a significant loss of product, especially in areas of high groundwater.

When a leaking tank is discovered, immediate safety issues are assessed and information is collected for site categorization, such as inventory records, precision testing records, and repair histories.

To categorize a site, a field TPH test is conducted in which TPH levels are measured in the ambient air, or air drawn from the soil. The result of the TPH test is compared to background levels. Background levels are determined by taking three soil samples from nearby or adjacent properties. If background TPH levels are exceeded, the site is placed in category II for further testing.

Under category II, quantitative lab analyses of soil are conducted for BTX and E, and TPH. Samples are collected from the bottom of the excavation at worst-case locations. The initial trigger levels for BTX and E are the detection limits of the laboratory procedure. According to the manual, the trigger levels should be in the order of 0.3 mg/kg. Cleanup levels for BTX and E are derived from precipitation rates and depth to groundwater, using tables detailed in the manual.

For TPH measurements in category II, a leaching potential analysis was developed to determine the levels of TPH that can be safely left in the soil. A leaching potential analysis is based on the tendency of TPH to migrate down to groundwater, depending on the features of the site. Four site characteristics which the manual considers important influences on migration of TPH include depth to groundwater, subsurface fractures, precipitation, and man-made conduits. Each characteristic is rated on a scale of high, medium, and low potential for leaching. These three degrees of sensitivity are expressed in terms of TPH that can be safely left in the soil, i.e., high (10 ppm), medium (100 ppm), and low (1000 ppm). The lowest sensitivity level determined for the four characteristics is used as a cleanup level. If a characteristic cannot be rated due to insufficient data, the lowest value (10 ppm) is used as a cleanup level. In addition, other site features may be considered, such as unique site characteristics, actual use of groundwater, and future land use.

If either the TPH levels or the BTX and E levels exceed the allowable limits, additional site analysis is needed. At this point in the evaluation process, the services of a registered geologist, engineer, or environmental chemist are recommended in the manual. In addition, the manual suggests that a general risk appraisal be conducted using environmental fate and chemistry data, and site-specific information. For this stage, computer modeling is used to estimate the concentration of BTX and E that can be left in place without risking groundwater pollution. The two models recommended for this step are the SESOIL model and the AT123D model. The SESOIL model involves long-term environmental fate simulation of pollutants which will enter groundwater. The AT123D model estimates the rate of pollutant transport and transformation in a groundwater system, and predicts groundwater contamination.

For category III sites having groundwater contamination, decisions of site investigations and cleanup measures must be made on a case-by-case basis. The step involves assessing groundwater use, and collecting and analyzing groundwater samples. The appendix of the field manual contains numerous procedures for field measurements of this type. Risk analyses for category III sites also require consultation with a regulatory agency or professionals in the field. The field manual does not attempt to offer guidelines for an in- depth risk assessment, and instead focuses on site categorization and laboratory analyses of indicator chemicals.

Fate Evaluation

Basis in Science

The LUFT manual is intended to provide practical guidance for dealing with leaking fuel tank problems. Its specific purposes are to provide assistance in investigating suspected and known leaks from underground tanks, assessing inherent risks, determining cleanup levels, and taking remedial actions.

The manual depends upon a general risk appraisal which relies upon generalized environmental fate and chemistry data and some site-specific information. This appraisal relies upon a simulated environmental system that was characterized by two computer models, SESOIL and AT123D. These models were utilized to estimate concentrations of BTX and E that can remain in soil without risking groundwater pollution.

The basis in science for SESOIL is presented in the chapter discussing SESOIL. The basis in science for AT123D is not presented in the LUFT manual and is not known at this time.

Applicability

The applicability of SESOIL is discussed in the chapter on SESOIL. The applicability of AT123D is not known. As stated in the SESOIL review, SESOIL can be applied to a wide variety of sites; however, the model requires substantial site-specific data for calibration. The site specificity of AT123D is not known.

Multimedia Relationships

SESOIL is structured for the integrated simulation of three cycles: the hydrologic cycle, the sediment cycle, and the dissolved pollutant cycle. Therefore, this model considers interactions among the soil-groundwater-surface and water-air compartments.

The ability of AT123D for use in the integrated simulation of environmental cycles is not known.

Input Data Requirements

SESOIL requires approximately 50 input variables, substantially fewer than required by many other models. Input variables can be classified into five categories: climate data, soil data, chemical data, initial chemical distribution in the soil, and washload data. However, it is most important to note that the general risk appraisal does not utilize most of these input variables.

The input requirements for AT123D are not known.

Strengths and Weaknesses

The strengths and weaknesses of SESOIL are listed in the chapter discussing SESOIL. The strengths and weaknesses of AT123D are unknown.

The weaknesses of the general risk appraisal, which utilizes SESOIL and AT123D, are:

- The appraisal only incorporates three types of site-specific data: precipitation, depth to groundwater, and extent of soil contamination. The appraisal, as a result, is overly simplistic and ignores important environmental fate processes.

- The manual does not state how the models were calibrated.
- The manual does not state if the models and the approach were validated.

Health Evaluation

Basis in Science

This is a state-prescribed decision-making method that uses limited field analyses in a "by the numbers" procedure to define the cleanup of soil contaminated by refined fuels.

It is based on the theory and practice of investigation of leaking underground fuel tanks and the actions to take based on guidance in the field manual. The approach taken goes beyond theory, since it incorporates field data. The general risk appraisal is not a true risk assessment, but it estimates the potential for contamination of groundwater. This tier of the field manual can utilize simulation models such as SESOIL and AT123D to assist in the appraisal.

Applicability

The manual is specifically intended to prevent groundwater contamination by gasoline and diesel fuels, with attention focused on benzene, toluene, xylenes, ethylbenzene and total petroleum hydrocarbon analyses. It indicates that lead and ethylene dibromide should be considered where appropriate. The manual is basically concerned only with soil contamination and potential for groundwater pollution with instructions for treatment or removal of soil based on comparison with worksheet values. Soil cleanup objectives are calculated in terms of the location and concentrations of these substances, and in terms of a small group of site characteristics. Primarily, groundwater criteria drive the cleanup objectives, although the manual was developed for the State of California soils. This approach should be easily transferred to other areas (e.g., other states).

The values are based on both drinking water action levels and other regulatory criteria. In addition, there are several decision points which are based on the results of a scoring system in response to questions and observations. The field manual, by design, is a general approach that can be easily applied to site-specific scenarios. These include factors such as precipitation, depth of groundwater table, and some geological parameters.

The manual addresses site contamination resulting from leaking underground storage tanks.

The manual differentiates sites on the basis of soil type.

Multimedia Relationships

The manual focuses on groundwater contamination.

Soil-groundwater-surface and water-air relationships are considered only indirectly, and only for soil/water via results of the aforementioned models (SESOIL and AT123D).

The guidance manual and specifically the two models described deal with the vadose zone and the groundwater zone separately, but can be linked so that results from the unsaturated zone can be used as the input for the groundwater model.

Input Data Requirements

The LUFT manual is not a computer model or a quantitative analysis.

Using the field manual requires, at the minimum, some data collection from a specific site so that at least there is a benchmark for comparison with the first tier in the decision tree. The simulation models do have other input data requirements, but many of these can be default values contained within the manual or can be easily obtained from references.

Specific Input Requirements:

- BTX and E concentrations in soil and possibly groundwater

- average annual precipitation
- distance from surface to groundwater
- estimated volume of contaminated soil

Strengths

The methodology is simple to apply, unambiguous, and comprehensive in the areas of applicability and should be useful for state employees. It also appears useful for emergency situations.

It is specific to underground storage tanks.

It is meant for petroleum products; it focuses on gasoline and diesel fuel products and therefore is directly applicable to most underground storage tank situations.

It recognizes that conditions vary from area to area and so it is a general approach that allows site-specific cleanup levels to be developed.

This field manual provides for a consistent approach in the appraisal of leaking fuel tanks in that the decision trees and the worksheets should lead the users to a similar determination.

The appendices are a particularly strong area of the document and should be especially useful.

Much of the manual also contains very pragmatic information on specific topics such as analytical techniques, sampling protocols, and judgmental statements, which can be extremely relevant in the field.

Weaknesses

The model does not address indirect pathways of exposure; it only considers groundwater and vapor hazard.

The simplistic methodology does not address all petroleum products or exposure routes. It deals with gasoline and diesel fuel products only. It does not deal with other hydrocarbon-based materials, such as waste oil or solvents. The model is also not very applicable to nonpetroleum problems.

Certainly, even for gasoline, some additional compounds should be included, i.e., manganese additives, methyl-t-butyl ether and (despite reasons given to exclude them) organo-lead compounds, ethylene dichloride, and ethylene dibromide. For heavier petroleum products and crude oil, polynuclear aromatic hydrocarbons would be important. The physico-chemical behavior of organic phase mixtures was evidently not considered.

Water quality limits are taken from Department of Health Services Sanitary Engineering Branch drinking water supply action levels. No detailed explanation of how these limits are derived is given.

Leaching potential analysis appears to be a very simplistic and inflexible way of determining acceptable levels of soil contamination. It needs more documentation of the data used to derive those values.

The computer models used in the general risk appraisal phase have not been validated.

The approach is limited to only a few consitutents of petroleum products. The model focuses only on BTX and E.

The model is a real "cookbook". Measure the soil contaminant levels, plug them into the models, look up the result on the health effects table, and find the acceptable level. This could also be considered a strength.

The discussions of detection limits (in ppm) and action levels on page 7 and page 88 (in ppb) present some confusion. Action levels on page 7 should be referenced. Some of the questions perhaps can be contained within a checklist. Field utilization of the manual has not been mentioned as to suitability of the guidance.

Comments

LUFT is more of a "rule of thumb" than a methodology. It is somewhat more sophisticated than Stokman and Dime[15] in that it allowed some site-specific discretion. However, there is no role for a biologically-based risk assessment. The only real issue to be determined at any site is whether or not the potential groundwater contamination by BTX and E is likely to be greater or less than predetermined limits.

GEOTOX - A Multimedia Compartment Model[11-14]

Developers

T.E. McKone and D.W. Layton

Available from

National Technical Information Service (NTIS)
General Information 703/487-4600

Phone

Sales: 703/487-4650

Request: Geotox Multimedia Compartment Model User's Guide May 1987
Project Order Number: 83PP3818

Description

GEOTOX is a set of programs used to calculate time-varying chemical concentrations in various environmental media. It uses a hypothetical model that includes continuous input of contaminants, dynamic distribution through a closed system, and loss via chemical and biological degradative pathways, with a steady-state solution that may, in some cases, reflect multiyear build-up to the steady state.

The program predicts movement of toxic substances in the environment and describes the partitioning of chemical species into air, water, soil, and sediments. The model also estimates human exposure to environmental contaminants. The output is a comparison of health risks.

GEOTOX is primarily a linked transport, transformation, and human exposure model. It employs steady-state equations to evaluate the transport and transformation of a contaminant among the eight environmental compartments, including atmospheric gas, atmospheric particles, biomass, upper soil, surface water, and lower soil. The manual shows the interconnections between these compartments in the three relevant media: air, water, and solids (soil, atmospheric particulates, and surface water sediments). These interconnections and consideration of the three media are comprehensive.

GEOTOX is described as a "global model." It was originally developed for ranking the potential health risks associated with toxic metals and radionuclides in the global environment. It has also been upgraded to evaluate organic chemicals, but the global nature of it remains. There are only four choices the user can make with regard to the geographic scenario setup: Northeast/Central U.S., Southeastern, Western, and California "eco-regions." Thus, GEOTOX lacks significant site-specificity application.

This global philosophy is also true for the eight compartments that are modeled. This being the case, GEOTOX must be considered the broadest of screening model methodologies for estimating exposures. The same lack of site specificity is true for the exposure algorithm. GEOTOX assumes a 70-year life span characterized by two phases: a 10-year childhood phase and a 60-year adult phase. The GEOTOX manual describes all relevant assumptions concerning ingestion, inhalation, absorption, etc., which appear to be based on recent and valid exposure literature. There are no algorithms in the model to evaluate health risk, i.e., no evaluation of exposures with regard to key health impact information for the modeled contaminants such as cancer potency factors.

Fate Evaluation

Basis in Science

GEOTOX relies upon partition coefficients which reflect the transfer of chemicals through the environment from soil or water to items consumed by humans. These partition coefficients are utilized to develop equations which describe the relationship between human intake and the concentration of a chemical in soil. Transfer rates between environmental compartments are estimated by utilizing a series of suggested equations. The output from these equations is utilized to conduct the risk assessment for the chemical and site in question.

The supporting documentation reviewed did not contain any data substantiating the validity of GEOTOX.

Applicability

The system is not limited to application to any specific soil type or dissolved organic chemical.

Multimedia Relationships

The output from each equation appears to be independent of the output of other equations in the approach.

Input Data Requirements

Supporting documentation for GEOTOX did not contain a summary list of input data requirements, although the documentation did identify them.

Strengths

As a result of its principal global orientation and lack of site specificity, this model offers no unique strength in assessing site-specific contamination.

Weaknesses

The first major weakness of GEOTOX is that the approach ignores basic principles and reactions governing chemical behavior in soil systems. The approach does not normally acknowledge most basic, known reactions and factors affecting reactions, such as:

- organic chemical reactions in soil (i.e., hydrolysis, general substitution, elimination, oxidation, reduction, soil-catalyzed reactions) although they can be specified.
- organic chemical biodegradation in soil factors affecting biodegradation rates, although they can be specified.

The second major weakness of this approach is the lack of validation.

The approach does not address free product in the following areas:

- depth of penetration of bulk hydrocarbons
- spread of free product
- migration rate of free product
- effect(s) of large concentrations of other organics on adsorption and mobility
- emission rate for a pure bulk hydrocarbon on a soil surface
- organic chemical biodegradation in soil factors affecting biodegradation rates

A problem with the exposure algorithms is that many assumptions are embedded in the model, and the user is unable to change them through data files.

Health Evaluation

Basis in Science

GEOTOX is a set of programs used to calculate time-varying chemical concentrations in various environmental media. The program predicts the movement of toxic substances in the environment and describes the partitioning of chemical species into air, water, soil, and sediments. The model also estimates human exposure to environmental contaminants.

Applicability

The model does not fully address a range of exposures applicable to petroleum contaminated sites. It does not specify industrial vs. residential scenarios; however, the description of the exposure parameters indicate primarily residential scenarios. An industrial scenario could, in theory, be used by modification of these parameters.

Only human (not environmental) health effects are addressed. The model is intended to simulate the environmental behavior of contaminants discharged steadily by industrial processes. The types of constituents of concern targeted for evaluation are residuals in military wastes (e.g., trace elements, radionuclides, and organic chemicals) in a system consisting of eight compartments. "The aim is ... to provide information on the characteristic behavior of each substance and the resulting exposure to individuals living their full life and receiving all of their air, water, and food from the contaminated landscape." Thus, the soil surface concentration of a volatile compound undergoes continual renewal without depletion over time. The model is presented as a risk management screening tool. It does not deal specifically with petroleum products or other mixtures.

The GEOTOX approach can be used to evaluate waste residuals from a number of industrial technologies. For purposes of screening, GEOTOX divides the continental U.S. into four eco-regions, which differ in ecological and climatological characteristics. These eco-regions are:

- Northeastern/Central U.S.
- Southeastern U.S.
- Western U.S.
- California

The methodology is not site-specific, as "these models use a representative landscape in order to describe the steady- state distribution of [chemicals] ... as a result of continuous additions to soil."

Typically site-specific issues relate to fate and transport, which do appear to be well addressed in this program (see Environmental Fate Evaluation section). However, assumptions on bioavailability do not appear to be adequate in this program. From the supporting material, it seemed as though an uptake of 100% (ingestion) was the default. This is clearly inappropriate in many situations.

The use of this model is very limited. As mentioned previously, its global structure eliminates its use for site-specific issues. Furthermore, although upgraded from an original version to handle neutral organic chemicals that are susceptible to degradation, it appears more valid for nondegrading (or extremely slowly degrading) contaminants such as the radionuclides and toxic elements (e.g., arsenic) for which it was originally intended. This assertion is made for two reasons: (1) that nondegrading contaminants and (in the case of toxic elements) contaminants that already exist in the environment, reach a steady state such that a broad geographic box-model approach is reasonable, and (2) perhaps most environmental problems with neutral organic chemicals are relatively small in size (Superfund sites, emergency spills, etc.) and well defined such that a site-specific model is far preferable to a global model. The one exception to this generalization might be pesticides, which are not small in size, but as a general and fair rule, degrade relatively quickly such that a broad, box-model approach is inappropriate.

Multimedia Relationships

GEOTOX is comprehensive in its multimedia considerations. The "source" term can originate from any of the eight compartments. As an example to describe the connections, consider the starting point as soil contamination. From there, the following routes and interconnections are considered for contaminant transport:

- in the soluble phase to lower soil and groundwater, and from groundwater to surface water
- in soluble and sorbed phase, directly to surface water; in the sorbed phase to bottom sediment of surface and back to the surface water
- in the sorbed phase through atmospheric particulates which can deposit on surface water and/or biomass (biomass reservoirs feed back to the soil phase)
- volatilization to the atmosphere, and from there back to the soil or to surface water

Exposure is estimated from reservoirs in each of the eight compartments, and exposure is estimated not only to humans, but to fish and cattle (because they are part of the human ingestion algorithm).

A very complex, and presumably complete set of multimedia relationships, couched in somewhat abstruse mathematical symbolism, is presented. Unfortunately, none of the documents provided on GEOTOX fully describes what is going on inside the computer, the equations used, or the sources of input values.

Input Data Requirements

The requirements to run the program approach are reasonable. The major input requirement is a source term of input of a chemical per unit area in terms of moles/km^2.

Physico-chemical properties and toxicity or carcinogenicity data are required for each contaminant. It appears that the model provides, for each such compound, a nominal rate of discharge into the environment of 1 mole/km^2/yr. Presumably, any other discharge rate would result in steady-state levels proportional to the scale-up (or scale-down) factor, and the overall exposure would be compared to the "virtually safe dose rate."

The GEOTOX model comes with 4 global environments and 14 chemicals, so if the user wishes to only consider these scenarios, there are no input data requirements. The user does, however, have to enter time step/source term data in the third input data set. The global environments are in 4 "LAND" files, and the chemical profiles are in 14 ".CHM" files.

Modeling different chemicals may not be difficult for the experienced exposure modeler. Several parameters are well understood and readily available (although not as part of this user guide). For example, familiar parameters include molecular weight, Henry's Law Constant, organic carbon partition coefficient, transformation (or decay) rate from the compartments, etc. Other familiar contaminant parameters are required which relate to exposure: bioconcentration factor in fish and meat (assumed equivalent), partition coefficient equaling concentration in beef meat and fat (meat and fat considered equivalent) divided by water, vegetation, and soil.

It may also be possible for the modeler to create his/her own global environment, in this case here it would probably be more difficult and less sophisticated. One reason is that there are several media-routing parameters for which no guidance is given in the manual, requiring the user to make his/her own judgements. Examples here include land surface runoff (cm/yr), irrigation from groundwater (cm/yr), deposition rate of suspended sediment (kg/m^2/yr), biota dry mass production (kg/km^2/yr), and so on. Other parameters require the user to have a better feel for the global philosophy of the model: biota dry mass inventory (kg/

km^2), fraction of the total surface area in surface water, groundwater inventory (kg/km^2), and so on.

It would seem that the intended user will not want to create a different global environment. He or she may wish to create a new chemical file, although the manual does not provide any background materials for appropriate parameter selection.

Strengths

The model is already built and field tested against reality. It also appears easy to alter.

The model is based on "first principles" and measured phenomenon and is well documented.

The program appears to be fairly complete for fate and transport considerations.

The multimedia compartment model can screen chemicals based on their inherent toxicity, persistence in the environment, and dilution. It has the ability to link established environmental concentrations with exposure pathways in order to project accumulated lifetime exposures within a population.

The model is computerized in such a way that the user has very little to do and can get results rapidly. The system could probably be revised so as to fit better in a spill or Superfund type of context. The differential equation format, which stresses multimedia relationships, may be highly desirable for thorough and systematic exposure assessment for various kinds of toxic releases.

The multimedia compartments and connections are comprehensive.

If it can be considered a strength, the intended model user will likely not need to define an environment of concern — a choice of four should be sufficient.

The experienced user can develop his or her own chemical parameter file, so there is flexibility from that perspective.

The model has the capability to handle a constant source term input/day or an input/day source term that varies over time.

The model also has a unique feature in tracking the fate of contaminants as they transform from one species to another. This was an option set up for radionuclides primarily, but may also be useful for organic chemicals. The manual is not clear as to how many transformation products can be handled.

The model can handle up to ten chemicals at a time if the user models only one constant source term for all ten (module GEOTOX-A), or four chemicals at a time if all four have varying source term inputs (module GEOTOX-B), or a combination of constant and varying source term inputs but only a maximum of four chemicals total (running both modules simultaneously, two with constant source term inputs and two with varying source term inputs, for example).

The portion of the manual describing how the model is actually run is clear. Input-output structure description is clear and examples are given. The graphical elements are useful, presenting contaminant concentration in individual media, uptake in terms of mg/km/day, and risk estimates.

One useful and unique component of GEOTOX is the differential sensitivity analysis. This analysis allows one to evaluate the influence of small changes in each input parameter on the final risk calculation. In addition, a Monte Carlo sensitivity analysis may be available in the future.

Weaknesses

It is unclear how user friendly the system might be or whether it would be useful for emergency settings.

It is not applicable to small contaminated sites, such as many petroleum sites would be.

The GEOTOX user manual needs a section which describes the scope, intended (and appropriate) use of the model. The section on design philosophy (page 5) is inadequate in describing the scope of intended use.

The manual needs a clearer explanation of the cellular structure of each compartment; i.e., it should be explicitly stated that each of the eight compartments is one large cell. If this structure were clear, then multimedia exposure modelers who require a more site-specific approach would know that this model is not an appropriate one for that use.

Along the same lines, the manual should be clear that the "ecosystem" structure and its four options on ecosystems, Northeast/Central, Southeastern, California, and Western, are the only choices the user has. It seems clear, in other words, that many of the ecosystem parameters are difficult to evaluate and probably require a greater understanding of the philosophical approach of GEOTOX, as well as appropriate differences among broad geographic regions.

Although the chemical parameters file could be uniquely generated by the user (given the chemical-specificity of the parameters as described earlier), the user guide contains no guidance on a general or specific level as to how a user could develop his or her own chemical file.

It should be made clear that all critical exposure assumptions like ingestion rates (for fish, meat, water, soil), inhalation rates, and so on, are not parameter input but rather are built into the model.

There are several uncertainties associated with the screening process. In general, there is a lack of data on various contaminants and parameters which affect their predicted risks. There is uncertainty in exposure due to the ingestion of crops and edible plants, little understanding of soil/plant partitioning, and an absence of information about environmental degradation and increased health risks.

There are several disadvantages to the use of this program with respect to the risk calculation component. Absorption was not adequately addressed for the different pathways. The exposure parameters presented in Figure 4 of the User's Guide Supplement appear reasonable as an average case, although it would be useful to have more detail relevant to specific regions of the country (e.g., vegetables or fish consumption). While it appears that the exposure period could be truncated for the risk calculations (to take into account different residential periods near a site), it was not clear how easily that could be performed.

Another disadvantage of this approach is the limited number of chemicals which exist in the library of the program. Two of the chemicals are relevant to petroleum contaminated soils, (specifically benzene and lead). Other chemicals of potential interest, such as xylene or toluene, do not have readily available data files.

The examples seem to present serious discrepancies. The loss rate constants for air and water seem far too high. Additionally, since all loss rate constants are of a similar order of magnitude, and distribution is primarily in the air and water compartments, it is hard to see why the time to reach steady state should take so much longer to attain for TNT and RDX than for benzene. No provision seems to have been made for losses off-site via transport through air or water. The equal sediment deposition and loss rates, approximately 25 cm/yr seem very high, and the meaning of these values is unclear. The implied aquifer thickness of 46 m also seems excessive. Equations 26 and 29 do not seem right.

The assumed dietary intake of plant matter is unreasonably high; for example, the number of people who grow and mill their own wheat, then produce and consume bakery products from it in the U.S. is surely infinitesimal.

Among the documents furnished, the description of GEOTOX is not sufficiently thorough for the user to understand what is going on in the computer program.

Finally, GEOTOX was not designed to handle risk assessment for crude petroleum, petroleum product mixtures, or used petroleum product mixtures.

Comments

From a global perspective GEOTOX would be very useful for evaluating fate and transport. It is quite comprehensive, considers different regions of the country, has some very attractive graphical features, and appears to be the most complete program so far produced with respect to multimedia relationships. However, there are some deficiencies with respect to the actual exposure and risk assessment part. The number of chemicals is limited, and site-specific bioavailability is not adequately addressed. In addition, the focus appears to be primarily on residential exposures and it is not clear how other types of exposures (e.g., industrial) could be incorporated in the program. Finally, it would be useful to know whether less than lifetime exposures could also be employed with this model. For the purposes of the Health Committee, with some modification and simplification, this model might well serve the needs of agencies to evaluate for petroleum contaminated soils.

Hawley's Assessment of Health Risk from Exposure to Contaminated Soil[16]

Developer

John K. Hawley

Available from

Open Literature

Description

A quantitative risk assessment methodology was developed based on analyses of contaminated surface soils, translated into absorbed daily doses, from which carcinogenic risks could be calculated. In addition to analytical data, the estimates were based on input values derived from the scientific literature, for example, the skin surface area of an adult likely to be coated with soil.

Hawley's assessment of health risks from exposure to contaminated soil addresses characterization of toxicity and the estimation of exposure. This assessment provides quantitative estimates of risk on the basis of measurement of soil contamination in a residential area (indoor and outdoor) from exposure via ingestion, inhalation, and dermal absorption.

This analysis is based on the assumption that contamination measurements are available only for soil. It is flexible and can be modified to incorporate measured values for contaminant levels in other media. The corresponding health risks associated with chemicals found in soil are estimated on the basis of assumed long-term exposures and dose-response relationships for chronic toxic effects of the contaminants.

Health Evaluation

Basis in Science

This assessment specifically involves risk from exposure to soil contaminants in residential areas (indoor soil/dust and outdoors) via inhalation, ingestion, and dermal absorption.

The author states that the results compare favorably with those obtained by the Center for Disease Control during their analysis of TCDD in Times Beach, Missouri.

Applicability

The approach treats only direct exposures (inhalation, ingestion, and dermal contact) to dust and soil, focusing particularly on dermal exposures. Intermedia transport is not addressed.

It appears that this approach can be applied to a variety of contaminants; however, different absorption fractions may be needed for petroleum products.

The approach is not site-specific. Variability of soil characteristics and its effect on absorption is not discussed, although it would probably be easy enough for users to plug in another value. Similarly, alternative climatological factors could be accounted for varying the exposure duration (fraction of week and fraction of year) components.

Introduction of such variables as land use and soil properties might certainly serve to modify the methodology.

Although the approach is not chemical specific, it has the capability of being made so.

Data are useful for site by site analysis.

Multimedia Relationships

The approach considers only exposures to contaminated soils. It does not consider exposure to contaminated water or air (except exposure to dust in the air).

The primary environmental pathway assessed in this analysis is the migration of contaminants from soil/dust to air.

The simple one-step pathways considered here involve no multimedia relationships.

Input Data Requirements

Soil concentrations of contaminants, along with the equations developed in Hawley's paper, are all the input needed. An estimated upper 95% confidence limit of the dose for increased cancer risk of 10^{-6} is also needed. It would be well for the user to reexamine the input needed, i.e., Hawley's basic assumptions, so as to determine how well they might apply to the specific compounds and sites being considered.

Default values include body weight and height (based on age), breathing rate, ingestion rate/day, absorption ratios and exposure duration (e.g., play hours/day/week) values.

Strengths

This assessment is very simple to use; it uses one equation. It gives very detailed explanations as to assumptions used to arrive at exposure level estimates.

This analysis is flexible and can be modified to incorporate measured values for other media, if available. Exposure estimates and risk assessment procedures are presented and can be used to estimate lifetime risks from contaminated soil in a residential area or by using an acceptable level of increased risk, to set a maximum contaminant level. In addition, this analysis takes into account indoor exposure. This exposure analysis is an extensive review of pertinent scientific literature and is a conservative approach to assessing health risks from soil contamination in a residential area.

The author thoroughly examined the direct exposure pathways from soil to humans and assembled most of the relevant available information. The procedure is easy to follow and not difficult to modify as additional information is developed.

Weaknesses

The approach has not been intended for petroleum contaminated sites.

Complex issues such as environmental half-life and decomposition products are not addressed. The approach does not account for contaminant degradation in soil.

No distinction is made between total body exposure and total body dose.

The approach needs to document the assumptions that 75% of inhaled particles is retained and that 100% of contamination of ingested or inhaled soil is absorbed.

The estimate of the quantity of dirt ingested by children (250 mg/day) may be high. More accurate estimates have appeared in the literature since the development of this methodology.

This analysis assumes a gastrointestinal absorption of 100%. This is not a realistic estimate for all soil contaminants. Dermal absorption of 100% is also not realistic.

The author provides the methodology to estimate exposure to humans living near a contaminated site in a residential setting. The only exposure pathways considered were inhalation of dust and soil, dermal contact with dust and soil, and ingestion of dust and soil. Inhalation of volatilized material, and exposure through other environmental media (i.e., contaminated groundwater) were not considered. The effect on exposure of fate and transport of contaminants was mentioned but not considered by the method. Future land uses were also not considered.

In the section on risk, the approach describes how to sum exposures to estimate an average lifetime dose. It is then suggested that a comparison of this value to levels at which chemicals might exert their toxic effects should be made. However, the author did not adequately identify sources for this information, and he did not describe how to estimate values of concern. The sections of the paper describing exposure estimation were substantially more detailed than those which discussed risk assessment.

It might be useful to consider the Hawley exposure estimation methodology in assessing the validity of assumptions in other models. However, the Hawley report by itself is of limited value for assessing the hazards of hydrocarbon contaminated soils. In particular, it should be noted that the method is "chemical specific" and does not attempt to describe the assessment of risk following exposure to a mixture.

Comments

The Hawley report was not intended to be a comprehensive model for the risk assessment of chemical contaminants in soil, and certainly not for petroleum product mixtures. It can serve as a module for a more encompassing assemblage of methods. The transdermal absorption of contaminants from soil needs much more research than is reflected in this report.

The scenario characterization assumes that soil ingestion peaks at 2 1/2 years, reaches its nadir at 6 years, and increases again to a constant adult rate. This is an unusual approach that seems to be derived by scaling from limited data.

Hawley's article has been widely cited by individuals completing analyses of soil-based exposures to contaminants, and it represents a good departure point for developing improved methods for this exposure pathway. His analysis of the dermal uptake pathway (uptake of a pure compound adjusted for matrix effects) is somewhat dated in light of recent experimental work with animals and fugacity models for estimating dermal uptake of chemicals with different physico-chemical properties. Furthermore, some of the parameter values presented need to be revised to reflect the results of recent studies.

New Jersey's Soil Cleanup Criteria for Selected Petroleum Products[15]

Developers

Sofia K. Stokeman and Richard Dime

Available from

Open Literature

Description

The New Jersey Department of Environmental Protection's "Soil Cleanup Criteria for Selected Petroleum Products" describes a soil cleanup methodology based on a few individual constituents of petroleum products which are believed to pose the greatest threat to public health. The most hazardous constituents were identified to be the carcinogenic polycyclic aromatic hydrocarbons (CaPAHs) and benzene. Acceptable soil contaminant levels (ASCL) were determined based on lifetime soil ingestion, a 1×10^{-6} cancer risk, and carcinogenic potency factors for individual constituents. The ASCL was compared to residual soil levels of CaPAHs and benzene resulting after soil cleanup to 100 ppm total petroleum hydrocarbons, as reported in the literature. For residual soil levels yielding a greater than 1×10^{-6} cancer risk, a lower soil cleanup level was proposed. With the exception of used motor oils from vehicles driven over 10,000 km, CaPAHs and benzene levels were below the concentration which would exceed a 1×10^{-6} cancer risk after cleanup to 100 ppm total petroleum hydrocarbons.

Health Evaluation

Basis in Science

The authors attempt to estimate the residual risks from carcinogens in refined and used refined petroleum products that remain in soil after cleanup actions. They focus on a cleanup level of 100 ppm for total petroleum hydrocarbons. The methodology is based on literature data regarding the content and carcinogenic potency in these products of benzene and carcinogenic polynuclear aromatic hydrocarbons, along with assumptions as to the rates of human ingestion of soils and of human inhalation of airborne particulates. Risk calculations are only presented for a "pica child" scenario. Concentrations in soil post-cleanup are obtained by reducing the initial product concentrations by a factor of 10^{-4} equal to the overall cleanup efficiency. EPA cancer potency values are used, with carcinogenic PAHs assumed to have potency equal to benzopyrene (BaP), and a 10E-6 individual risk level is assumed negligible.

Applicability

The authors intend the methodology to be broadly applicable to soil sites contaminated by petroleum where contact by children is a concern. It is noted that groundwater concerns are not addressed.

This methodology might be applied to defining product-specific surface soil criteria for various refined or used refined petroleum products, based on the two surface exposure pathways.

The method is not site-specific.

Multimedia Relationships

The approach focuses on chronic effects due to long-term exposures to contaminated soils. Multimedia relationships, including the evaporation of surficial benzene, are not addressed.

Only soil contact by children is addressed.

Input Data Requirements

The approach requires the concentration of contaminants in soil and the estimated upper 95% confidence limit of the dose for increased cancer risk of 10^{-6}.

If the approach were to be applied for risk assessment, only the petroleum product's maximum content of benzene and carcinogenic polynuclear aromatic hydrocarbons, along with their carcinogenic potency factors (now called "slope factors") would be needed, and these are found in the article under discussion.

Strengths

The approach focuses on petroleum products; it attempts to deal with risk analysis of complex mixtures. It is simplistic and implementable.

Weaknesses

The approach focuses only on carcinogenic risk. It needs to compare the cleanup levels calculated when considering carcinogenicity to those involved with other toxic endpoints.

It assumes the only toxic component is PAHs. For certain materials, noncarcinogenic endpoints may be important and causative material nonPAHs. The API study on carcinogenicity of various fractions of crude oil show total aromatics do not always correlate with carcinogenic potential. Carcinogenic potency factors are all assumed to equal that of BaP.

Only a single surface exposure pathway, ingestion, is of concern (particulate inhalation being a negligible component), and only the carcinogenic petroleum components are evaluated. The model does not address other exposure pathways, such as ingestion of contaminated groundwater. The exposure assumptions are, in the main, far too conservative. For example, surface evaporation of benzene is ignored, polynuclear aromatic hydrocarbons are assumed to be completely absorbed, and soil ingestion is estimated at 2.5 g/day. A more accurate estimate of quantity of dirt ingested by children is available in the literature.

The calculation of risks for a "pica" child needs attention. In addition to the discussion concerning an appropriate ingestion estimate (above), the details of the calculation to translate this rate into a lifetime average rate are not given. There appears to be an error. The annual grams per day should be calculated first and then averaged over a lifetime. This calculation would reflect a changing body weight, and would not use a constant of 70 kg as it appears was done here. This would affect results appreciably.

It is not clear from the information presented that benzene and

PAHs are the compounds of dominant concern for petroleum contaminated soils. Some attempt to compare the toxicity of whole petroleum products to the toxicity of these constituents is needed. Further examination of the literature may reveal other toxic species of concern. Metals in used oils are not mentioned.

Exposure from some petroleum products will change over time. One would not expect that the composition of aged spilled petroleum would resemble the initial product. Volatiles would be lost and nonvolatiles would concentrate. This would be expected to change the conclusions drawn for benzene and PAH's (in opposite directions).

Comments

This paper deals explicitly with the potential health risks of petroleum contaminated soils. Although it represents a starting point for assessing the health risks of petroleum products, it does not integrate information on the transport and transformation of the hydrocarbons and potential exposure pathways.

Draft Interim Guidance for Disposal Site Risk Characterization in Support of the Massachusetts Contingency Plan[17]

Developer

Massachusetts Department of Environmental Quality Engineering Office of Research and Standard

Available from	*Phone*
Massachusetts State Bookstore State House Room 116 Boston, MA 02133	617/727-2834

Description

This draft guidance document is notably consistent with the guidance provided in EPA's Superfund Public Health Evaluation Manual. It describes four methods of risk assessment which may be applicable to a particular disposal site. These risk assessment methodologies include the following parameters: hazard identification, exposure assessment, standards, sets of cleanup levels, risk characterization, guidelines, policies, and dose-response assessment.

Health Evaluation

Basis in Science

Generally, standard exposure and risk equations are used. A set of default dermal absorption fractions are given for various chemical classes. The scientific basis for these is very weak. Exposure to infants via mother's milk is explicitly included, a pathway commonly not considered. However, no procedure is given for predicting contaminant levels in milk. The procedure does not present or specify the fate and transport models.

Extensive use is made of references to Massachusetts regulatory documents and databases, but there is a strong component of quantitative risk assessment (with little detailed guidance) involving individual short pathways, rather than scenarios. Physicochemical relationships are of almost no concern, so that heavy reliance is put on analytical data (as opposed to models).

Applicability

The primary materials evaluated are termed oil and hazardous materials (OHM). One of the proposed methodologies would apply to disposal sites at which public health standards exist for each OHM reported in each contaminated material. Another method would apply to disposal sites for which there are established cleanup values and there are additional methods which would apply to single medium and multimedia instances when there are no established cleanup values or public health guidelines.

The methodology is intended to be applicable to all situations. It does this by being relatively inflexible and relying on assumptions and default values wherever possible.

Emphasis is on human health effects endangerment assessment of individual chemicals, summed as total hazard indices and carcinogenic risks. The guidance does not address petroleum product mixtures.

The procedure is not readily applicable to determining cleanup levels because it is not linked to fate and transport models. The procedure could be used to estimate exposures and risks for a wide variety of sites and chemicals, where monitoring data or modeled concentrations are available. While some chemical-specific toxicity data is included, no data on chemical/physical properties was provided.

The methodology is intended to be applicable to all sites in Massachusetts. It does this by apparently assuming that all sites are equivalent at least in terms of geologic and meteorological characteristics.

There is some degree of site-specificity built into the draft guidance document. For instance, the determination of background levels will be site-specific. The evaluation of the demographics around the site also contributes to a site- specific approach. For instance, potential human receptors may be identified based on proximity to the site (residence), nearby worker populations, or children and/or trespassers on the site if the site is accessible. The potential for future exposures is also evaluated based on future land use.

Virtually no guidance is given for incorporating site-specific information, nor are the models developed to cover cases where exposure is strongly site-dependent. Reliance is placed on general assumptions and exposure-point analytical data. (Averaging periods would, however, be site-related).

Multimedia Relationships

The methodology lists soil, surface water, groundwater, and air as environmental media of concern. It does not provide a method of estimating contaminant distribution into the various compartments. Rather, it seems to assume that contaminant concentrations will be known from direct sampling and that, unless there is data to the contrary, these concentrations will never change.

No detailed procedures are included for considering multimedia transfers. However, the procedure mentions that volatiles can be emitted from domestic water and cause exposures comparable to drinking water.

Input Data Requirements

The method requires an inventory of mean and maximum concentrations of all contaminants in all relevant media (soil, groundwater, surface water, and air), knowledge of receptors, and a listing of acceptable regulatory levels for the various contaminants.

Primary input requirements are exposure parameters such as ingestion rates and exposure durations. Default values for many of these are provided.

Other specific input requirements include:

- exposure concentrations for each contaminant to which humans may be exposed
- for each contaminant, information concerning toxicokinetics, mechanisms of actions, toxic effects (including dose response data), and interactive effects
- identification of potential receptors
- identification of exposure points, routes, durations, and patterns
- bioavailability constants
- since no computer program is associated with this document, all data must be provided by the user, in particular receptor-related values and toxicological criteria, in addition to analytical data

Stengths

The methodology is straightforward and requires relatively little data input other than contaminant levels.

Stress on use of analytical data minimizes validation problems. Some recent references, cited in Appendix B of the document, seem worth pursuing and utilizing elsewhere. Where input data is available, the calculations are relatively simple. Mother's milk exposure is a useful, if frequently overlooked, pathway.

Guidelines for hazard and exposure assessment appear to be very detailed concerning what items need to be incorporated into a risk assessment and how exposures are calculated.

Weaknesses

The model's focus is on individual compounds and not on complex mixtures.

Cancer potency factors are taken from the IRIS database. This is a nonpeer reviewed database. Recent outside reviews have shown numerous errors.

It does not allow for use of carcinogenic potency factors derived by different models.

There should be a sliding scale of 10^{-4} to 10^{-6} and not just 10^{-5} for acceptable levels of exposure to carcinogenic risks.

The basis for selecting a hazard index of 0.2 as a threshold for cleanup is inadequately documented.

While the draft clearly documents all the elements which should be considered in a risk assessment, it implies that these considerations are all straightforward and simple. In fact, the approach reduces the decision analysis to comparing two calculated values. Very little consideration is given to detailing how these often subjective decisions are made.

The system does not seem to make specific use of IARC classification to determine if the material poses a risk to humans.

The major disadvantages of this document are the uncertainties that result from the use of such parameters as exposure duration and ingestion rate estimates, as well as the assessment of exposures and toxicities of environmental chemical mixtures.

The methodology is inflexible, not site-specific, does not incorporate fate and transport models, bases exposure estimates on arithmetic rather than geometric means, and does not justify its assumptions.

The use of models is so minimal that there is over-dependence on analytical data. No statistical methodology is presented, however, for dealing with the data. In fact, the document generally lacks detailed guidance of any sort. Some exposure pathways are missing, e.g., soil to air, shower water to air (except via one reference) and food chain. The significance of physico-chemical properties is ignored, so that no time-related projections are feasible. Most assumptions are not qualified as regards the nature of the compounds or receptors. Appendix A of the document applies a 20% factor that is evidently based on EPA's "Relative Source Contribution", but fails to explain or justify it. This document depends far too heavily on references to Massachusetts source documents for it to be used outside that state.

MYGRT: An IBM Personal Computer Code for Simulating Solute Migration in Groundwater[18]

Developer
Electric Power Research Institute

Available from
Electric Power Research Institute
3412 Hillview Avenue
Palo Alto, CA 943304

Phone
415/855-2411

Project RP2485-1
EPRI Project Managers: Ishwar P. Murarka and Dave A. McIntosh
Energy Analysis and Environment Division
Contractor: Tetra Tech, Inc.

Basis in Science

MYGRT is a system comprised of a one-dimensional code and a two-dimensional code. These codes are used to predict concentrations of an organic chemical in groundwater downgradient of a disposal site as a function of time or distance. The primary flow direction can be horizontal or vertical. The environmental processes accounted for in the codes include advection, dispersion, and retardation.

Applicability

MYGRT is designed to address chemicals in the saturated zone.

The model is not limited to application to any specific soil type or dissolved organic chemical.

MYGRT can be applied to a wide variety of sites, and it does not require substantial site-specific data for calibration.

Multimedia Relationships

MYGRT is not structured for integrated simulation of other environmental compartments. This model does not consider interactions among the soil-groundwater-surface water-air compartments.

Input Data Requirements

Use of the model requires over twenty input parameters.

Strengths

MYGRT does not require a user who is relatively "sophisticated" or input from databases containing substantial quantities of data. It does not require the use of a computer possessing capabilities greater than an IBM-PC compatible.

The model should be applicable on a wide variety of sites because it does not require substantial site-specific data for calibration.

Weaknesses

MYGRT is not a well-recognized or well-utilized model that is readily accepted by that segment of the scientific community which utilizes soil-chemical fate models.

No data and information have been presented to show that the model has been extensively validated and shown to work under a number of scenarios.

The model cannot accommodate the physical, chemical, and biological reactions of the chemical under scrutiny in soil systems. These reactions include:

- organic chemical reactions in soil (i.e., hydrolysis, general substitution, elimination, oxidation, reduction, soil-catalyzed reactions)

- organic chemical biodegradation in soil
- organic chemical adsorption

The model cannot be utilized on sites with very large vertical variations in soil properties.

The model addresses dissolved organic chemical movement in the saturated zone. Regarding free product, however, the model does not address:

- depth of penetration of bulk hydrocarbons
- spread of free product
- migration of free product
- effect(s) of large concentrations of other organics on adsorption and mobility
- emission rate for a pure bulk hydrocarbon on a soil surface

The model is limited in its ability to produce exposure data and information that can directly be utilized during the risk assessment process.

Pesticide Root Zone Model (PRZM)[19-20]

Developer

USEPA Lab, Athens, Georgia

Available from

Center for Exposure
Assessment Modeling
c/o U.S. EPA
Contact: Ms. Greene
College Station Road
Athens, GA 30613

Phone

404/546-3549

Facsimile (FAX)

404/546-3340

Description

The Pesticide Root Zone Model (PRZM) is a one-dimensional dynamic, compartmental model for use in simulating chemical movement within and below the plant root zone. Predictions can be made either daily, monthly, or annually.

The output of PRZM is a time series of chemical leachate mass and concentration leaving the root zone and entering the water table.

Fate Evaluation

Basis in Science

PRZM has two main components: hydrology and chemical transport. The hydrology component, which simulates runoff and erosion, is based on both the Soil Conservation Service curve number technique and the Universal Soil Loss Equation. Evapotranspiration (from soil and plant) is estimated by on-site pan evaporation or is calculated from air temperature and the number of daylight hours when pan data is unavailable. Water movement is simulated by a model based on soil-water capacity parameters including field capacity, wilting point, and saturation. Runoff from snow melt and daily storm erosion can also be accounted for by PRZM.

The chemical transport component models chemical uptake by plants, surface runoff, erosion, decay/transformation, vertical movement, foliar loss, dispersion, and retardation. Degradation of the chemical is represented by a single first order process with the rate coefficient specified by the user for each defined zone. Thus, all biotransformation and decay processes are represented. Different rates can be used to account for the processes occurring in different layers of the soil profile.

The user of PRZM performs dynamic simulations of chemicals applied to soil and foliage. This type of modeling allows the consideration of pulse loads, the prediction of peak events and the estimation of time varying pollutant profiles, thus overcoming the limitations of more steady-state models. The model estimates the vertical movement of contaminants.

The soil profile is divided into a number of layers. For each soil profile, the model solves the solute transport equation which includes the processes of advection, dispersion, adsorption, degradation, and plant uptake. For the soil surface layer, additional terms are included into the equation to account for surface runoff and erosion. Soil parameters can be specified separately for each zone. Each zone is divided into uniform layers. As many soil layers as are necessary to accurately represent the soil profile can be utilized.

PRZM has undergone a moderate amount of field testing, primarily to simulate pesticide leaching. Both the EPA Office of Pesticide Programs (OPP) and chemical manufacturers have used PRZM to assess the potential leaching of new and currently registered pesticides. OPP has also used PRZM to develop a method for expeditiously screening pesticides for potential groundwater contamination. Results of testing have been consistently positive when comparing average field values to PRZM predictions of soil concentrations and mass flux to groundwater.

Applicability

This model is designed to address chemicals in the unsaturated zone. It can be applied to generic environmental conditions for the purpose of evaluating the general behavior of a chemical. The model is not limited to application to any specific soil type or dissolved organic chemical.

Multimedia Relationships

PRZM has been coupled to groundwater, surface water, and air models for assessing the effects of tillage on pesticide concentrations in soil. Therefore, this model can be used to assess interactions among the soil-groundwater-surface water-air compartments. PRZM considers runoff and its effect on contaminant transport.

Input Data Requirements

Use of the model requires a number of input variables, but appears to require substantially less input than that required by many other models. Much of the required input is available in the literature, at least for a general problem.

PRZM has 41 data input requirements, 11 related to hydrocarbon transport. While most of the input parameters could be collected from the literature, the total number of site-specific data inputs could be as high as 34. Few of these inputs are typically available from a site assessment.

Stengths

The model is well recognized and accepted by that segment of the scientific community which utilizes fate models. The model has been validated and shown to work under some scenarios.

The model can accommodate the physical, chemical, and biological reactions of the chemical under scrutiny in soil systems. These reactions include:

- organic chemical reactions in soil (i.e., hydrolysis, general substitution, elimination, oxidation, and reduction)

- organic chemical biodegradation in soil
- organic chemical adsorption

The model is compartmental, allowing for a significant amount of user tailoring of the model to a specific data set or site conditions.

Weaknesses

The model requires a user who is relatively "sophisticated" and input from databases containing substantial quantities of data.

The model requires the use of a computer possessing capabilities greater than an IBM-PC compatible.

For input variables not readily available to the model user, very complex calculations will be required to obtain this data. These calculations and the decisions made to select the appropriate equations for use will require a most sophisticated user.

The model addresses dissolved organic chemical movement in the unsaturated zone. Regarding free product, however, the model does not address:

- depth of penetration of bulk hydrocarbons
- spread of free product
- migration rate of free product
- effect(s) of large concentrations of other organics on adsorption and mobility
- emission rate for a pure bulk hydrocarbon on a soil surface

Comments

Early versions of PRZM apparently had significant limitations which have been documented in the literature. The material reviewed is for Release II, which is supposedly modified to overcome those limitations. PRZM results have been compared to field data and were shown to predict concentrations within a factor of two. As can be seen by the above description, PRZM handles many processes especially related to plants which are not of interest in dealing with surface oil spills. It would appear that these processes could be set to zero or eliminated, and the model could be used to deal with problems of interest.

While the model appears to be technically valid for the conditions for which it was developed, it is not *directly* applicable to simulating hydrocarbon transport in the unsaturated zone from a surficial leak or spill.

The model contains a number of processes which are not directly related to hydrocarbon transport such as plant up-take, active treatment zone, etc. While the user could possibly ignore the nonhydrocarbon related factors with potentially little loss of accuracy, the resultant model would be a hybrid and potentially viewed as less credible than the original model.

PRZM, VADOFT (Vadose Zone Flow and Transport Model), and SAFTMOD (Saturated Zone Flow and Transport Model) are three subordinate models that comprise the Risk of Unsaturated/Saturated Transport and Transformation of Chemical Concentrations (RUSTIC) model. RUSTIC predicts pesticide fate and transport through the crop root zone, unsaturated zone, and saturated zone to drinking water wells. Only the PRZM was considered relevant and therefore reviewed by CHESS; further reference to RUSTIC is found in: U.S. EPA 1989, "Risk of Unsaturated/Saturated Transport and Transformation of Chemical Concentrations (RUSTIC), Vol. I: Theory and Code Verification," EPA/600/3-89/048a.

POSSM

PCB On-Site Spill Approach[22,23]

Developers

S.M. Brown and A. Silvers

Available from	*Phone*
Electric Power Research Institute (EPRI) Research Reports Center Box 50490 Palo Alto, CA 94303	415/965-4081

Description

POSSM is a contaminant transport model developed to predict environmental concentrations associated with a chemical spill. The model predicts daily changes in chemical concentrations on a spill site's soil and vegetation as well as losses of chemical due to volatilization, surface runoff/soil erosion, and leaching to groundwater. The model was originally developed for PCB spills.

Fate Evaluation

Basis in Science

POSSM is a modification of the Pesticide Root Zone Model (PRZM). It is a one-dimensional, compartmental, dynamic, transport, and fate model. It is one-dimensional because water and chemical movement in the soil are assumed to occur only in the vertical direction. It is compartmental because the soil column can be divided into a number of layers, each possessing different soil properties. The model is dynamic because a daily time step is utilized to calculate changes in hydrologic conditions and chemical concentrations.

Applicability

POSSM is designed to address chemicals in the unsaturated zone. The model is not limited to application to any specific soil type or dissolved organic chemical.

POSSM was written for small surface spills and may not be appropriate for underground tank leaks.

POSSM contains simulation strategies that can treat petroleum as a single component, groups of components, and soluble-insoluble fraction. The model does consider equilibrium relationships with the insoluble fraction of a spill.

POSSM can be applied to a wide variety of sites, and does not require substantial site-specific data for calibration.

The spill area, spill volume, and depth of penetration are considered. The model also has the capability to predict emission fluxes from impervious surfaces.

Site-specific data include soil bulk density, soil field capacity, soil wilting capacity, soil organic carbon content, and depth to groundwater.

Multimedia Relationships

POSSM is apparently structured so that its output can be utilized for integrated simulation of other environmental compartments.

POSSM calculates chemical concentrations due to leaching to the groundwater, volatilization, runoff, and erosion.

In addition, POSSM contains three submodels that predict changes in environmental concentrations away from the spill site:

- The PTDIS Gaussian plume dispersion model estimates airborne concentrations at any distance downwind of the source.
- The RIVLAK contaminant transport model can estimate chemical concentrations in rivers or lakes.
- The GROUND contaminant transport model can estimate concentrations in a saturated groundwater system.

Therefore, POSSM has the capability of predicting the concentrations in the air, in surface waters and in the groundwater. POSSM also calculates concentrations in soil, vegetation, and impervious surfaces.

Input Data Requirements

Use of the model requires approximately 31 input parameters.

1. Chemical:
   - adsorption coefficient
   - degradation rate
   - boiling point
   - molecular weight
   - solubility
   - dispersion coefficient in air
   - Henry's Law Constant
2. Soil Characteristics:
   - bulk density
   - organic matter content
   - field capacity
   - wilting point
   - depth to groundwater
3. Climate (daily):
   - precipitation
   - pan evaporation
   - air temperature
   - wind speed and height measurement
   - solar radiation
4. Spill and soil surface:
   - spill area
   - slope
   - average runoff duration

- soil erodibility
- vegetation type
- rooting depth
- cover density
- date of spill
- spill quantity
- depth of penetration
- extent of soil, vegetation and impervious surface cleanup

5. Impervious Surface:
   - surface
   - depression storage depth
   - solids loading rate
   - Efficiency and frequency of street cleaning

Strengths

It appears that POSSM does not require a user who is relatively "sophisticated".

The model should be applicable on a wide variety of sites because it does not require substantial site-specific data for calibration.

The model mathematically considers important environmental processes including percolation, infiltration, runoff, evapotranspiration, and volatilization. It also considers degradation; however, one or more degradation processes must be expressed as a single first-order degradation rate constant.

The immiscible phase is considered only in equilibrium partitioning with the aqueous, solid, and vapor phases.

In addition to the air dispersion, groundwater, and surface water model, POSSM contains a risk assessment model, EXPOSE. EXPOSE calculates potential exposure levels, inhalation, ingestion, and dermal contact.

POSSM is based on PRZM which has been tested for pesticide fate and is recommended by the EPA.

Weaknesses

POSSM is not a well-recognized or well-utilized model that is readily accepted by that segment of the scientific community which utilizes soil-chemical fate models.

No data and information have been presented to show that the model has been extensively validated and shown to work under a number of scenarios.

The majority of the references cited in the supporting texts contain data, information, and approaches that are not current state-of-the-science.

The model cannot accommodate, on an individual basis, the physical, chemical, and biological reactions of the chemical under scrutiny in soil systems.

The model may require the use of a computer possessing capabilities greater than an IBM-PC compatible.

The model addresses dissolved organic chemical movement in the unsaturated and saturated zones. Regarding free product, however, the model does not address:

- depth of penetration of bulk hydrocarbons
- spread of free product
- migration rate of free product
- effect(s) of large concentrations of other organics on adsorption and mobility

The model also does not consider volatilization fluxes when water is evaporating; this is a factor that has been found to affect the amount of chemical that can enter the atmosphere, significantly depending upon the chemical. If the chemical has a high Henry's Law constant, like benzene, water evaporation is not significant, but for chemicals with low Henry's Law constants, like benzopyrene and chrysene, it is important. Compounds with low Henry's Law constants ($< 2 \times 10^{-5}$) have been found to have insignificant volatilization without water evaporation, but their surface concentration can build up under evaporation and volatilization may become significant.

The PRZM model upon which POSSM is based inaccurately

accounted for the volatilization of metalaxyl. Carsel et al.[20] admitted in their analysis that there were volatilization losses that were unaccounted for in their model. It does not appear that this has been adjusted for in POSSM.

The model assumes that the spill is stationary and that no redistribution of the oil occurs with time.

POSSM does not consider organic matter in the transport calculation. The organic matter content is input and appears to be used only for the erosion calculation. Organic matter has been found to be extremely important in aqueous soil solutions in the chemical partition of nonionic organic compounds. The omission of the organic matter content for calculation of adsorption is a weakness of the model since it is important in determining the mobility and also the partitioning between the vapor phase and chemical adsorption. Modifications of Kd by dividing it by the fraction of organic matter has been shown to reduce the coefficient of variation of adsorption significantly. This would be a simple modification to the model.

The 1988 POSSM Manual from EPRI does not contain the MCPOSSM with the Monte Carlo simulations, which would be very useful in modeling the environmental fate of hydrocarbons.

The input data requirements do not specify the viscosity of the MAPL phase, which would be very important in the transport and fate of the hydrocarbons and individual constituents.

PPLV

Preliminary Pollutant Limit Value Approach[24,25]

Developers

Mitchell J. Small and David Rosenblatt*

Available from

The report which documents this approach is Technical Report 8918 (July 1988) as well as computer software developed under Work Unit 686 (Hazard Assessment Method Computerization) in the Environmental Quality Research Branch of the Health Effects Research Division.

Description

The primary goal of the PPLV approach is to estimate allowable concentrations of a contaminant in different environmental media. A basic premise of the PPLV approach is that the chemicals of concern are at or near equilibrium in soils, air, and water. Based on this assumption, partition factors (measured or estimated) are used to determine the distribution of a contaminant in media and foods.

* The Defense Technical Information Center does not supply software. The author will supply the software if you provide either two unmarked two-sided 5-1/4 inch floppy disks or one unmarked 3-1/2 inch diskette and a suitable mailer. The floppy disks may be either double-density or high-density. Send disks to:

Commander
U.S. Army Biomedical Research and Development Laboratory
ATTN: SGRD-UBG-E (Mr. Small)
Fort Detrick
Frederick, MD 21702-5010

Preliminary Pollutant Limit Values (PPLVs) are acceptable concentrations of a pollutant at a site. PPLVs are derived from Single Pathway PPLVs (SPPPLVs) for a specific contaminant. SPPPLVs represent linear exposure pathways from the site to the human receptor (e.g., soil-water-human or soil-plant-human, etc.) SPPPLVs are based upon the acceptable daily dose of the pollutant (DT), partition coefficients, standard intake factors, and a body weight factor.

The hazard posed by a specific contaminant is represented by DT, which is based on already established standards, allowable daily intake (ADI), maximum contaminant level (MCL), or threshold limit values (TLV), or derived from the no effect level (NEL) or LD_{50} from animal studies. The exposure level is described using partition coefficients which may include octanol-water partition coefficient (K_{OW}), soil organic carbon adsorption coefficient (K_{OC}), water solubility, saturation vapor density, and others. All SPPPLVs which present critical exposure pathways are factored into the calculation of PPLVs to determine the cleanup level for a specific contaminant.

The method focuses primarily on issues relating to fate and transport of all pollutants from soil. By considering 7 routes for water and 13 routes for soil contact, equations are developed to relate the concentration of a chemical in soil to its concentration in water (e.g., beef or vegetables) or air to estimate human exposures. The risk is then assessed through the use of what are essentially reference doses (RfDs) or cancer potency factors.

Much of the input relates to parameters to estimate the relationship between the concentration of the contaminant in soil and the relevant medium for a particular scenario. For exposure parameters such as soil ingestion or inhalation rate, standard factors are proposed. For toxicity factors, the reader is referred to various EPA manuals to obtain either RfDs or cancer potency factors.

The PPLV approach is based in theory, but has been used in dealing with contaminated sites at several Department of Defense installations. The author discusses three sites: the Rocky Mountain Arsenal, the Alabama Army Ammunition Plant, and a landfill 30 miles west of Saginaw, Michigan, the Gratiot County landfill. At all three sites highly persistent chemicals such as the polybrominated biphenyls were present. These types of chemicals would not be directly applicable to petroleum products.

Fate Evaluation

Basis In Science

The PPLV approach relies upon partition coefficients that reflect the transfer of chemicals through the environment from soil or water to items consumed by humans. These partition coefficients are utilized to develop equations that describe the relationship between human intake and the concentration of a chemical in soil. The output from these equations is utilized to conduct the risk assessment for the chemical and site in question.

The supporting documentation reviewed did not contain any real data substantiating the validity of the PPLV approach.

Applicability

The system is not limited to application to any specific soil type or dissolved organic chemical.

Multimedia Relationships

The output from each equation is independent of the output of other equations in the approach.

Input Data Requirements

The supporting documentation for the PPLV approach did not contain a summary list of input data requirements, although the documentation did identify them.

Strengths

There are no general strengths to this approach.

Weaknesses

The primary weakness of the PPLV approach is the inability of the approach to generate concentration estimates that normally are utilized to conduct risk assessments. The approach relies solely upon partition coefficients, which in and of themselves have limited applicability in adequately characterizing chemical behavior in soil systems.

The second major weakness of the PPLV approach is that the approach ignores basic principles and reactions governing chemical behavior in soil systems. The approach does not acknowledge even the most basic, known reactions and factors affecting reactions, such as:

- organic chemical reactions in soil (i.e., hydrolysis, general substitution, elimination, oxidation, reduction, and soil-catalyzed reactions)
- organic chemical biodegradation in soil
- factors affecting biodegradation rates

The third major weakness of this approach is the lack of validation.

The approach does not address free product in the following areas:

- depth of penetration of bulk hydrocarbons
- spread of free product
- migration rate of free product
- effect(s) of large concentrations of other organics on adsorption and mobility
- emission rate for a pure bulk hydrocarbon on a soil surface
- organic chemical biodegradation in soil
- factors affecting biodegradation rates

Health Evaluation

Basis In Science

The primary goal of the PPLV approach is to estimate allowable concentrations of a contaminant in different environmental media. A basic premise of the PPLV approach is that the chemicals of concern are at or near equilibrium in soils, air, and water. Based on this assumption, partition factors (measured or estimated) are used to determine the distribution of a contaminant in media and foods.

The PPLV model approach can be utilized to determine analytical levels for investigative survey sampling requirements. The PPLV approach should be useful for other environmental hazard assessment methodologies and incorporates reasonable treatment of toxicological data and pathways for human exposure into a computational framework, whereby acceptable environmental contaminant levels may be derived. It is a conceptual framework incorporating much current environmental thinking. The approach provides a useful framework in which to construct answers to environmental problems and from which to define needs for further study.

The PPLV approach is comparable to other procedures for calculating risk assessment through several pathways. Scenarios composed of individual pathways are mathematically modeled. Compound-specific values are provided from databases or are estimated, largely from regressions based on the properties of similar compounds. Site- and receptor-specific data is required as input.

In the document, a methodology of ranking toxic effects is described in case information on a NOEL (no observed effect level) is not available for a particular chemical. This approach seems reasonable; however, there are drawbacks. Some users of this approach would have insufficient knowledge in toxicology in order to appropriately interpret the effects observed in animals. Many toxicological endpoints are not listed in the document. The pertinent considerations and exposure scenarios are discussed. The report does not provide an integrated model approach to evaluate a site.

Applicability

The PPLV approach has been applied to multiple sites with a variety of contaminants with varying physical properties. Multiple methods used to calculate the SPPPLV make it applicable to a variety of sites.

This approach is presented with all equations and can be used for sites other than those presented in the document. The matrix approach for identifying multipathway considerations is especially useful in such applications.

The PPLV approach offers flexibility because it considers a range of pathways and multiple scenarios. Thus, different developmental scenarios could be readily evaluated for a single site and compared. However, PPLV seems to be constrained by the requirement for equilibrium conditions and like other models, it becomes more difficult to use as the contaminant of interest becomes more complex. For example, the source of contaminant at a site, such as benzene from a gasoline spill, is not infinite. The source may be depleted, there may be biodegradation, and so forth. All these phenomena will contribute to non-equilibrium conditions and it is not clear whether such half-life issues are addressed in the methodology.

The PPLV approach could be used by state agencies; however, some exposure parameters need to be updated.

The emphasis is on the human health effects of individual chemicals, with back-calculation to acceptable cleanup levels. The methodology was not developed with mixtures, let alone petroleum products, in mind.

The PPLV approach was utilized in the evaluation of soil contamination specifically for three sites: the Gratiot County Landfill, the Alabama Army Ammunition Plant, and the Rocky Mountain Arsenal. Various land use scenarios were evaluated (current or projected activities) such as farming, housing, and industry. Established partition coefficients, bioconcentration factors, and other default values for specific chemicals are also identified and utilized in the proposed exposure assessments. Some of the constituents evaluated were polybrominated biphenyls, explosive related chemicals, lead residuals, pesticides, chemical warfare agents, and various other organic and inorganic substances.

The site specificity is quite detailed and describes in general terms all of the factors that need to be considered. Despite provision of default values, the methodology is intended to be as site-specific as possible, and users are encouraged to tailor each operation to the case under consideration.

PPLV seems applicable to a wide variety of sites. It can incorporate site-specific information (and provides sources for such information), and can consider certain portions of a site separately. The approach is flexible enough to handle several water- and soil-based exposure pathways that may be possible at a site.

Much of site specificity is a fate and transport issue. However, one critical issue of site specificity with respect to toxicity is the influence of the soil medium on bioavailability. This is particularly important with high molecular weight hydrophobic compounds such as the PAHs. It may also be an issue with respect to volatile compounds such as benzene. There are no bioavailability components associated with the toxicity factors. Thus, from a toxicological perspective, the approach inadequately addressed site specifity.

In addition, site specificity is important with metals, particularly with lead, where the bioavailability of lead compounds can vary greatly according to the intrinsic solubility of the compound and the particle size in which the compounds are associated. Thus, lead would not be addressed appropriately with this methodology.

The PPLV approach uses the "Si" factor (dilution factor) in f(Ki, Si) that describes ease of transmission of a pollutant from soil, which is site-specific information about soil.

It can accommodate cases of food contamination, such as beef, where only a portion of the animal's food is contaminated.

The approach emphasizes the need to be site-specific, and details current use expectations.

Multimedia Relationships

All environmental compartments and intercompartmental exchanges are considered.

There are several multimedia interactions discussed through the PPLV model. In particular, this model addresses the migration of contaminants in water, soil, and sediment to biota (e.g., aquatic life, domestic animals, and plants). PPLV provides methods of estimating soil/groundwater, soil/surface water, and soil/air partitioning. It also provides methods to estimate partitioning from groundwater or surface water into biological media.

Most of this is based on default assumptions. Some of the assumptions are not relevant for petroleum products since these compounds do not bioaccumulate to any great extent.

The PPLV approach considers all pathways of exposure, including contaminanted foodstuffs. It was not clear, however, whether there was a parameter for evaluating runoff from soil into receiving waters.

Input Data Requirements

Input data consists of the physico-chemical properties of the organic chemical being assessed (e.g., Henry's Law Constant, vapor pressure, molecular weight, organic carbon partition coefficient, etc.) and environmental properties (e.g., soil porosity, organic carbon content of soil, etc.).

The user will have to provide much of the data, though a database for a variety of compounds of interest to the Army is gradually being assembled. The usual types of input are required: "acceptable" dose levels or toxicity data leading thereto, physicochemical properties of contaminants and of soil, demographic information, anticipated land uses (for choice of scenarios), and meteorological data.

Several default values are used for each pathway. There should be a better effort to standardize these default values as they are being used in various agencies, the EPA, for example.

Requirements:

- Exposure related variables. These are given as default values. The assumptions and reasons for selection are provided and justified.
- Accumulation of toxicants in meat and milk. As above, default values are provided and justified.

- Geologic and Climatic Factors. A range of real values is provided, default values are suggested, and sources for additional information are suggested.
- Physical/chemical properties. Real values can be used. Alternatively, methods of estimation are provided.

Specific input requirements:

- Standardized intake factors based on body weights and rate of pollutant uptake.
- Partition coefficient.
- Acceptable dose of a contaminant, Dt.

Strengths

The methodology is pragmatic and reasonable. It has a good, thorough approach and seems generally straightforward and easy to understand. At the same time, it provides realistic estimations of exposure.

Good documentation of default values is provided along with sources of information for site- or region-specific information.

A computerized version exists.

The approach does advise testing PPLV values to determine if they are meaningful.

It is flexible regarding routes of exposure, discrete steps in each exposure route, and portion of pathway contaminated; determines the percentage of dose due to various land uses and routes. Good analysis is provided of indirect pathways.

It has been used on several sites, a variety of routes and land use scenarios.

SPPPLV equations are available for a variety of exposure routes.

It does note that standardized intake values should be adjusted if necessary.

It contains a thorough discussion of each factor, how it was derived, and the strengths and weaknesses of use.

One of the primary strengths of this approach is the identification and use of multimedia relationships. Contrary to many exposure assessment models, the PPLV approach avoids the use of complex mathematical equations, if the available data do not support them. It provides a useful framework in which to construct answers to environmental problems and from which to define a need for further study.

The methodology covers most of the exposure pathways that would be encountered at a given site. The calculational procedures, including methods of estimating selected properties and partition factors that might not be available on a chemical of concern are carefully documented. Although a formal uncertainty analysis has not been performed using this approach, an analysis showing how the different exposure pathways change as the log-octanol water partition coefficient increases is provided.

Essentially PPLV is a "big picture" approach and appears to provide a fairly full description of the site. However, some major reservations exist, both technical and editorial, that are addressed in the next section.

Weaknesses

The document does not guide the reader through a specific approach. However, all of the parameters are given and the reader with experience in this area would be able to calculate cleanup levels. However, using the provided information the cleanup levels might not be relevant for petroleum products.

Partition coefficients on a complex mixture such as petroleum products may be impossible to obtain. Partition coefficients of individual components will not reflect behavior of the mixture. It would be difficult to model a multicomponent mixture.

PPLV requires the assumption of an equilibrium state; it cannot be used to model spills.

PPLV apparently assumes that contaminants will not migrate away from the site. Thus, estimations of long-term exposure will be overly conservative.

Like AERIS, PPLV makes some conservative assumptions, including 100% bioavailability following inhalation or ingestion, and partitioning into plants from groundwater.

Absorption following dermal contact is ignored.

It would be worthwhile to examine whether uptake from plants and/or meat really represents a serious concern. This might be one way to simplify the methodology.

The magnitude of safety factor should be determined on a case-by-case basis, instead of utilizing standardized safety factors.

SPPPLV equations would be more useful if use of distributions of values for such things as average daily intakes, body weights, partition coefficients, and bulk densities of soils was allowed.

Assumption of strict additivity when calculating hazard from multiple materials may result in unreasonably low levels for individual components.

The PPLV approach does not take into account such human exposure pathways as dermal absorption of contaminants in groundwater or surface water, as well as the inhalation of volatile contaminants from groundwater/surface water. In general, this model utilizes default values for specific chemicals and is limited by theoretical simulations as opposed to real site-specific data.

It is not pointed out that, for certain groups of chemicals, bioaccumulation is not of specific concern. At high concentrations of the chemical in soil, plants may not grow; thus, consumption of contaminated produce would not be a problem. Some chemicals are rapidly metabolized and excreted by animals and levels in meat and milk would be no higher than in drinking water, or might even be less. This is not explained. There is no discussion of the extent of the contamination and how that would impact on the approach taken. At what point would an extensive cleanup be unnecessary?

The bioavailability problem is not addressed. No allowance has been made for less than 100% bioavailability or dust retention.

Breakdown and microbial degradation of the contaminants in the environment are not discussed. The relationship between concentration of a chemical and degradation is ignored.

Lifetime exposures are assumed to be at an equilibrium when in reality the exposures would decrease. Weather conditions, soil conditions, drainage, etc., are not considered.

One of the major weaknesses of this document is that it is quite poorly written. On page 37 there is another example of how the text in this document can be quite confusing. In the footnote, the document describes a 10-kg child and the amount of water ingested by such a child and then contrasts that with the intake by a

10-year old child. Page 25 provides a good example of how confusing the terminology in this document can be. For example, "NT" refers to the number of animals with tumors at the lower selected dose group. Apparently "T" refers to treatment, not tumors, which is apparent from the term "NC" which refers to the number of animals with tumors in the control group. The use of such confusing terminology makes this document very difficult to use. The use of abbreviations is also confusing. The same letter appears to mean different things depending on whether it is lower case or upper case. In general, the approach would be very difficult for a junior individual at a state regulatory agency to both use and understand. In addition, there are some significant concerns about several of the exposure parameters and about the way the dose issue is addressed.

A list of documents on page 15 of the PPLV document are described as resources for obtaining DT values. There is no clear evaluation of which documents are the best, nor is the IRIS database developed by EPA even noted. Perhaps the best source for EPA toxicity factors initially is the IRIS database and then the Health Assessment documents that are prepared by the Environmental Criteria and Assessment Office. These documents are not even noted. In addition, the ATSDR toxicity profiles are a useful resource and should also be reviewed initially, especially considering that the ATSDR documents are fairly current.

The document notes that the Health Advisories developed by the Office of Drinking Water consider that 80% of the intake of a pollutant is from sources other than drinking water. While this is correct and the actual background exposure will vary from pollutant to pollutant, it is odd that the PPLV report does not have a consideration for background exposure. This is a particularly important issue for noncarcinogens, such as lead.

Calculation of soil ingestion exposure basically averages the daily exposure over a lifetime. In fact, the dose rate will be much greater with young children than with adults. While the document notes this, it does not sufficiently address its importance, which is an issue for both carcinogens and noncarcinogens. For noncarcinogens, a daily dose which exceeds the RfD could occur in childhood, yet with averaging over a lifetime no exceedance of the RfD may be noted.

In addition, for carcinogens, age-specific risks must be developed. The risks should then be added to yield a lifetime total risk. This is an issue that would be particularly important for chemicals that are metabolized, such as PAHs. For chemicals with long half-lives that are poorly metabolized, such as PCBs or dioxins, which are not likely contaminants of petroleum products, this issue becomes less important.

A methodology is developed on page 21 of the PPLV document, which utilizes the TLVs as a way for estimating Dt. This approach is not recommended. The TLVs consider different effects from acute irritancy to cancer and chronic ailments. Sometimes the susceptible population for ambient exposures may be found among the worker population (e.g., exercising individuals exposed to ozone) and sometimes the susceptible population is not found within the worker population (e.g., young children exposed to lead). The bottom line is that the background documents to which the TLVs refer can be very useful in developing exposure limits. However, the blanket use of the TLV with uncertainty factors and dose adjustment is not warranted.

The document notes on page 23 that bioassay tests, although similar to chronic toxicity studies, are not designed to estimate a No Observed Adverse Effect Levels (NOAELs). While this may be correct, in fact, the number of animals that are involved in bioassay tests and the extent of the pathology actually make bioassay tests fairly appropriate for use in development of NOAELs. An uncertainty factor of ten is presented as a way of developing a NOAEL from a Lowest Observed Adverse Effect Level (LOAEL). An uncertainty factor is suggested for application to death or pronounced life shortening. This represents a misuse of the concept of LOAEL. The LOAEL should be considered a level at which effects of relatively minimal or moderate severity occur. Such effects might include fatty liver or mild liver necrosis.

On pages 27 and 28 the relationship between contaminants in water and fish are described as bioconcentration factors. More appropriately, bioaccumulation factors should be used which consider the contribution to the fish of contaminants in sediment, not just the dissolved concentration in water. This is particularly important for dioxin; however, it may also be important for other contaminants found in petroleum wastes, such as PAHs.

On page 36 the document notes that ingestion of water in an extremely hot region could be as much as 10 l/day, which sounds quite high. In Taiwan, for example, among farmers in a fairly hot region, EPA has estimated each individual's average water intake as about 3 l/day.

With respect to fish ingestion, any methodology must be sufficiently flexible to consider a range of fishing uses, including the sports fisherman and, where appropriate, ingestion factors for subsistence fishermen. These ingestion rates for subsistence fishermen could be very high — on the order of 90 to 100 g/day.

Risk assessment for contaminants in milk should, to the extent possible, consider the fat content of the milk that is ingested by individuals. Many individuals ingest lowfat or 1% fat milk. This could have a significant impact on hydrophobic contaminants which may be contaminating milk.

Incidental soil ingestion values are based on a hypothetical study by Hawley,[25a] which went into a great deal of detail, considering the minimal amount of data employed. Soil ingestion rates for children independent of soil pica should also be considered.

The choice of the geometric mean or the arithmetic mean as the appropriate descriptor depends upon the information one is seeking. If one is trying to estimate the total cancer burden predicted in the population, the arithmetic mean is the more appropriate statistic. However, for an evaluation of a risk of a particular individual, a geometric mean is a more appropriate statistic. For the ingestion uptake/biokinetic model, the soil ingestion rate that should be employed is the ingestion rate that is best calibrated to the model.

On page 45 there is a discussion of inhalation of windblown dust. The document assumes a situation with dust concentrations as high as 10 mg/m^{-3}, all of which are respirable. This is not consistent with what is known about the size distribution of soil contaminants in airborne particulate matter. In fact, soil contaminants are underrepresented in inhaled particulates. In addition, the assumption that all soil particles are from the site is erroneous.

On page 67, the document notes that only carcinogens that are expected to cause similar cancer should be included in a summation. The basis for this assumption is unclear since the issue is the total cancer risk from a site.

User's Manual for Risk Assessment/Fate and Transport Modeling System (1989)[26]

Developer

Scientific Services Section,
Division of Special Investigations

Available from	*Phone*
Commonwealth of Pennsylvania Department of Environmental Resources Office of Special Investigations 18th FL. Fulton Building Harrisburg, PA 17105	717/783-9475

Description

The Risk Assessment/Fate and Transport (RAFT) Modeling System was developed to provide Pennsylvania Department of Environmental Protection risk managers with a scientifically defensible and consistent basis for making decisions regarding actions taken to protect the public health from organic chemicals at waste sites and facilities.

The model includes a variety of equations that describe the movement of a chemical in various parts of the environment such as surface soil, saturated zone, unsaturated zone, air, etc. The objective of these equations is to predict the theoretical exposure people will receive from chemicals at a site. The model also includes equations to determine the risks associated with the computed exposures.

The model consists of ten separate Lotus spreadsheets, seven of which are fate and transport models. The model requires an IBM-PC compatible computer with a copy of Lotus 1-2-3 and a user with an elementary knowledge of Lotus 1-2-3. The movement of chemicals in a specific portion of the environment is computed by a single worksheet.

Each spreadsheet requires from 5 to 30 entries by the user. The entries include chemical-, site- , and population-specific parameters and depend on the portion of the environment being modeled. Examples of inputs for the Deep Soil Contamination Depletion Model spreadsheet include fraction of organic carbon in soil, K_{oc}, depth of contaminated soil, etc. The spreadsheet program computes output values using these inputs and the equations in the spreadsheet. The outputs from the deep soil contamination model are "soil contaminant concentration" and "average contaminant concentration in infiltrating water and duration of time at that concentration". The former is then used in the spreadsheet that computes the risk the site poses and the latter is used in the spreadsheet that models movement in the unsaturated zone.

The RAFT modeling system is designed to translate measured or predicted concentrations of a contaminant in different environmental media into dose rates via different individual exposure pathways. The calculated dose rates for noncarcinogens are compared with acceptable intakes by computing the ratio of estimated intake to safe intake (i.e., a hazard index). For carcinogens, a lifetime, incremental risk of cancer is calculated by multiplying a cancer potency factor by a lifetime average intake. The RAFT system consists of several modules that simulate the time-varying and/or steady-state movement of a contaminant in surface soil, a subsurface soil, and the saturated zone.

Most of the spreadsheets relate to fate and transport models and calculate the concentration of a soil-derived contaminant in a particular medium over time. Concentrations can be predicted for air, water, soil, and vegetables. In addition, an exposure calculation is available for mothers' milk. A separate module is available for contaminated drinking water. In some instances it is necessary to provide real data that, in other models, could be derived from soil or water concentrations (so that the cleanup levels cannot be calculated). Compound-specific values will presumably be provided. Site-specific data is required as input.

In this modeling system interpolations are based on Henry's Law Constant and the organic carbon partitioning constant for soil (surface soil and deeper soils), runoff, and evaporation are considered.

Vegetable contaminant concentrations are based on water solubility and on sorption characteristics of the particular chemicals. Contamination of leaves is based on the volatilization characteristics of the contaminants.

Fate Evaluation

Basis In Science

This system contains a number of individual models used in deriving cleanup standards. The system is designed to address chemicals in the unsaturated zone, the saturated zone, and the volatilization of chemicals from soil. The entire system is composed of ten Lotus 1-2-3 spreadsheets. Seven of the ten are exclusively devoted to fate and transport models.

The Shallow Soil Contaminant Depletion Model is used for shallow contaminated surface soils; it predicts how long a chemical will exist in the soil, how much will be released from the soil to the atmosphere, and how much of the chemical will be solubilized into percolating soil water that is migrating through the unsaturated zone toward groundwater or bedrock.

This model yields three values: (a) an annual soil chemical concentration, (b) an annual emission flux, and (c) an average chemical concentration in percolating soil water.

The Deep Soil Contaminant Depletion Model is used for deep contaminated soils. It predicts how long a chemical will exist in the soil and how much of the chemical will be solubilized into percolating soil water which is migrating through the unsaturated zone toward groundwater or bedrock. This model yields two values: (a) an annual soil chemical concentration, and (b) an average chemical concentration in percolating soil water and the period of time the chemical remains at that concentration.

The Air Release from Soils Model predicts chemical concentrations in the atmosphere due to chemical volatilization from surface soils by first computing a contaminant emission flux rate and then using an atmospheric dispersion box model.

The Unsaturated and Saturated Zone Contaminant Migration Model and the Unsaturated/Saturated Zone Coupling Model are utilized to predict migration of a chemical through the unsaturated and saturated zones.

The Soil Loss Estimation Model utilizes the Universal Soil Loss Equation to predict annual soil transport via overland flow for relatively small watersheds.

The mathematical treatment of those transport pathways is based largely on other published studies and appears to be technically sound. However, the draft user's manual does not provide enough information on the actual implementation of the models, apparently because the emphasis of the report is on the operation of spreadsheets that make up the RAFT system.

Applicability

The system is not limited to application to any specific soil type or dissolved organic chemical. Although the model can be applied to a wide variety of sites, the models within the system require some site-specific data for utilization.

Multimedia Relationships

The output from each model can be utilized as input to some of the other models in the system. The following mechanisms can be modeled: (1) on release to the soils, (2) contaminant uptake by plants, and (3) unsaturated/saturated zone "coupling".

Input Data Requirements

The input requirements are appropriate for this type of model and are not excessive or complex. The input requirements include data which are not typically available but are provided by the model as "default" values. The data input is straightforward and well-defined by the user's manual.

The system requires the following input parameters:

- aqueous solubility of each chemical
- concentration of each chemical in soil
- concentration of each chemical in groundwater leaving the site

- degradation rate constant of the chemical
- depth of contaminated soil
- depth of soil to bedrock or groundwater
- dispersion coefficient
- distance to receptor well or point of compliance
- duration of discharge of the chemical
- duration of emission
- effective air diffusivity of each chemical
- effective porosity of saturated zone
- fraction of organic carbon in soil (unsaturated and saturated zones)
- height of an imaginary box above the contaminated area in which an individual may be exposed
- Henry's Law constant
- linear distance perpendicular to wind direction
- molecular weight of each chemical
- organic carbon partition coefficient for each chemical
- percent slope of land
- saturated zone volumetric water content
- slope length
- soil bulk density
- surface area of contamination
- thickness of groundwater mixing zone
- time period that chemical will be discharging from under the site toward receptor or point of compliance
- total soil porosity
- unsaturated zone volumetric flux
- unsaturated zone volumetric water content
- vapor pressure for each chemical
- wind velocity

Strengths

The input requirements are not excessive, and default values are provided. The model appears to be quite easy to use, in a Lotus spreadsheet format.

Weaknesses

The primary weakness of the models comprising the system is the model's and system's dependence on an overly simplistic approach, which ignores basic principles and reactions governing chemical behavior in soil systems.

The second major weakness of this model is the lack of validation data. The output values are very difficult to verify in the field.

The model addresses dissolved organic chemical movement. Regarding free product, however, the model does not address:

- depth of penetration of bulk hydrocarbons
- spread of free product
- migration rate of free product
- effect(s) of large concentrations of other organics on adsorption and mobility
- emission rate for a pure bulk hydrocarbon on a soil surface
- organic chemical biodegradation in soil
- factors affecting biodegradation rates

The model has not undergone extensive peer review.

Comments

The individual spreadsheets are not interconnected, but could be incorporated into an expert system, also, some of the units used are confusing.

The model utilizes a "pulse" contaminant source. A contaminant mass depletion approach (e.g., exponential decay term) is physically more realistic. The contaminant source definition in this model leads to inaccuracies in the unsaturated/saturated coupling module. The "pulse" source will result in unrealistically high concentrations at the receptor well. The model should incorporate mass in the contaminant source term (not simply concentration).

Health Evaluation

Basis In Science

The RAFT modeling system produces a site-specific multipathway assessment of risk posed by waste in a single location. The exposure scenarios include nine ingestion pathways (soil, water, fish, sediment, milk, meat, vegetables, indoor dust, and mother's milk by an infant), three dermal pathways (soil, sediment, and water) and three inhalation pathways (outdoor vapors, indoor vapors, and outdoor particulates).

RAFT attempts to determine acceptable levels of organic contaminants and some inorganic contaminants in soil based on fate and transport modeling, exposure assessments, and risk assessments. Scenarios evaluate shallow and deep soil contaminant depletion models, air release from soil, and unsaturated and saturated zone contaminant migration. Default values for molecular weight, soil porosity, and other chemical-specific parameters are utilized in the RAFT approach.

With regard to fate and transport algorithms, it is fair to say that the simplified, steady-state equations employed are consistent with other steady-state modeling. However, there are other steady-state models with fewer simplifications, not to mention a wealth of complex, transient models that are an order of magnitude more realistic but also two orders of magnitude more difficult to use. While the authors acknowledge the simplicity of their approach, particularly in the groundwater modeling section, they also lose sight of it. An often repeated phrase is, "the answers will only be as valid as the values used as input. Ensuring that reasonable and valid input data are used is extremely crucial to the validity of any output from this spreadsheet." It is also true that the answers obtained will only be as valid as the gross assumptions and simplifications that go into the equations.

The document should be credited for some key features in the fate and transport modeling. It employs a time step of one year and update reservoirs of contaminants at the end of each year. The media spreadsheets (soil, groundwater, air) are run sequentially if the output from one feeds into the next. The models perform mass balance and reality checks at critical junctures; partitioning into the water phase is only allowed up to the solubility of the chemical, and chemical reservoirs cannot be dissipated beyond what's there to begin with, and so on.

Applicability

The RAFT system is site-specific to the extent that measured contaminant concentrations at a site (e.g., concentration of a contaminant in soil) are used as input to the soil modules along with other physical parameters of the site environment (e.g., soil porosity, bulk density, etc.). However, the basic structure of the modules is fixed, and it appears that it would have to be changed for sites that are different. Also, the present system is intended for single substances at low environmental concentrations. The movement of pure compounds or mixtures is not addressed. Hence, an analysis of a petroleum product or products in soil using RAFT would have to be carried out on a substance-by-substance basis.

The model was developed to evaluate HSCA/CERCLA sites, RCRA sites, emergency spills, and other waste management sites — in general any scenario where the starting point is soil contamination. RAFT was developed as a means of assessing the hazards of contaminated sites in Pennsylvania. Consequently, some of the geologic and meteorologic data, as well as some default assumptions (i.e., number of days in which the average temperature exceeds 32°F), are specific for Pennsylvania. Presumably these could be changed if other site-specific information was available.

There is also a stand-alone indoor air contaminant volatilization algorithm called TAPWATER which predicts concentrations of contaminants that volatilize from tap water considering all common household activities: showering, dishwashing, and so on. This stand-alone algorithm also estimates exposure and risk from these concentrations. Essentially any contaminant can be modeled as long as the user supplies critical characteristics of that contaminant: Henry's Law Constant, molecular weight, organic carbon partition coefficient, solubility, vapor pressure, and a background concentration and depth of contamination.

The RAFT approach outlines its exposure assessment model by illustrating its applicability to soils of wood-preserving facilities, inactive transformer manufacturing facilities, soils of an auto repair shop, and a garden. The types of constituents evaluated are pentachlorophenol, xylene, polyaromatic hydrocarbons, aroclor, trichoroethene, toluene, *p*-chlorobenzene, tetrachloroethene, phenol, naphthalene, and chloroform.

Applicability of the RAFT system is general and open-ended, requiring the user to have extensive knowledge of the site, receptor populations, and chemical properties. Output appears to be entire spreadsheets. It is difficult to know what assumptions were made in order to obtain the outputs of the model. Therefore, to some extent the model is a "black box" subject to abuse in that it will produce risk/cleanup numbers with no explanation. However, in the right hands it could be a valuable and apparently valid means of computing exposure and risk.

RAFT appears most able to accept site-specific data, but the basic system, as defined by a series of spreadsheets, is fixed.

Much of the site-specificity issue relates to the fate and transport models. However, the bioavailability factor (which would be site-specific for many petroleum contaminants) can be modified by the user.

It is possible to enter measured concentrations in various media or estimate those concentrations using the models. The final computations are only as accurate as the parameters entered and/or as accurate as the fit of the model to the site. The models selected for review appear to be state-of-the-art.

In the same way that any contaminant can be modeled, any site can be modeled with appropriate soil, groundwater, air, and other fate and transport parameters. The manual provides more detailed information for some key parameters specific to Pennsylvania conditions.

Although the RAFT approach takes into account several land uses, a variety of organic and inorganic chemicals, and considers off-site exposure, the types of sites for evaluation of contaminated soils are not exhaustive. Landfills and mining sites are not discussed, but may be incorporated if this approach is further developed. Site-specific factors may be utilized with the use of default values for each contaminant (if available). Since the exposure assessments of this approach are not based on real sampling data, the model is based solely on theory and simulations.

Multimedia Relationships

The basic system structure follows a multimedia approach. For example, a volatile contaminant in a surface soil is exhaled to the atmosphere and then atmospheric concentrations are predicted. Similarly, a contaminant can leach to the unsaturated zone and then to groundwater (saturated zone). It is not clear, however, if mass is conserved over all the compartments.

RAFT is capable of estimating the transfer of contaminants between soil and groundwater, soil and air, and soil and surface water. The methodology also considers loss from the surface by erosion. In estimating transfer to groundwater, the methodology considers transport through soil and water solubility. Transfer into the air is based primarily on volatility and assumed to occur only from the upper 2ft of contaminated soils.

The relationship between a contaminant in soil and windblown particles did not appear to be addressed in this method. Windblown dusts need to be evaluated, particularly with respect to metal contaminants and high molecular weight organic contaminants. In addition, the modeling of a contaminant from soil to plant did not consider contaminant uptake into leaves, but only contaminants entering the leaves directly from transpiration from air.

RAFT seems to deal with other multimedia relationships in an acceptable manner, except for (a) the soil-to-sediment-to-fish pathway, (b) the soil-to-cow-to-milk pathway, (c) the soil-to-cow-to-meat pathway, and (d) soil-to-vegetable routes. The factors used to calculate the uptake via the above-mentioned pathways are presented but the equations are not.

RAFT very systematically requires the user to begin evaluation by running a shallow (<2 ft) or deep (>2 ft) soil depletion model, depending on his assessment of where the soil is contaminated.

The shallow model feeds into the air volatilization model — volatilization is assumed negligible if the contamination begins at greater than 2 ft. Both the shallow and deep models feed into an unsaturated zone transport model which feeds into a ground water model. The volatilization model feeds into a vegetable model (deposition and adsorption in leaves). There is no surface water component in RAFT, although the authors indicate that the soil loss model can feed into a surface water algorithm. The soil loss model's purpose in RAFT is to estimate dissipation from the site via this route (only if contamination begins at the soil surface and the modeler chooses to run the shallow soil model.

Besides surface water modeling, there are two other missing and potentially important considerations. The first is root uptake of contaminant from the top soil models (shallow and deep) into the vegetable model. As written, there is no link, but the model requires a soil water concentration which is derived from contaminant sorption characteristics and soil characteristics. In estimating soil water concentrations in this way, the model neglects depletion of the contaminant reservoir through leaching, volatilization, and soil loss. It may be true that plant uptake of contaminant through the routes is such a small fraction of total dissipation and/or such a small fraction of total plant accumulation of contaminant (given the equations they used), that neglecting depletion of a soil reservoir may be acceptable. However, if this were the case, it should have been acknowledged in the manual. The second missing consideration is the transport of contaminant from the site via airborne particulates. As stated earlier, the indoor air volatilization model is a stand-alone product.

Input Data Requirements

Numerous equations, and some receptor-specific values are provided. Properties of the contaminants must be obtained elsewhere (in contrast to the old Superfund Public Health Evaluation Manual). Guidance toward some good sources is provided. Virtually nothing is said about the estimation of physico-chemical or toxicological properties. There is no accompanying computer software package.

The model requires the concentration of contaminant in the

relevant medium, information on ingestion, inhalation or dermal contact rate, absorption factors, body weight of exposed individual, and toxicity factors for subchronic, chronic, and carcinogenic effects.

To fully assess soil depletion, the required inputs include:

- Henry's Law Constants
- fraction organic carbon in soil
- organic carbon partition coefficients
- aqueous solubility of each contaminant
- total soil porosity
- effective air diffusivity
- soil bulk density
- unsaturated zone volumetric flux
- unsaturated zone volumetric water content
- molecular weight
- depth of contaminant in soil
- concentration of contaminant.

In addition, release to air requires duration of emission.

Migration in unsaturated and saturated zones requires:

- pore water velocity
- dispersion coefficient
- depth of soil to bedrock or groundwater
- degradation rate constant of contaminant
- duration of discharge

Potential food contamination requires the log octanol/water partition coefficient.

Finally, the risk assessment models require knowledge of the amount of exposure by different routes, length of exposure, body weight, and, if comparisons are to be made, the allowable daily intakes.

RAFT allows for a maximum of flexibility by requiring the user to input all necessary parameters for the fate and transport algorithms, contaminant description, exposure assumptions, and risk calculations. The manual also provides good summary information on valid parameter input for all algorithms with three exceptions: the water fate and transport and air algorithms; critical contaminant health endpoints (cancer potency factors and concentrations for noncarcinogenic effects); and fate and transport parameters for contaminants. There are, however, two "reserved" chapters (not yet completed) which are purported to contain these variables — an "Input Variable Values" chapter, and an appendix on "Selected Parameter Values for Common Contaminants". The input data requirements seem to be already available for petroleum compounds but are not provided.

Strengths

The methodology addresses the salient factors governing the movement of organic contaminants originating in surface and subsurface soils. It explicitly requires measured soil concentrations as input, which provides a direct link to site characterization data. RAFT specifically addresses most of the principal exposure pathways associated with hazardous waste sites.

The RAFT model and this supporting manual appear to be very useful screening tools for evaluating the impact and risk to exposed individuals from soil-contaminated sites, and separately from volatilization of contaminants from tap water in an indoor setting. The manual is clear and concise. Lotus 1-2-3 "cells" are explicitly described as either input cells or output cells. When it is necessary to use output from one algorithm as input to another, this is also clearly explained. The nine ingestion scenarios, three dermal contact scenarios, and three inhalation scenarios of the risk estimation models are a very reasonable menu of exposure options. Two other key strengths of the manual are that it provides many examples at the end of each chapter and within an appendix, and that it provides good information on reasonable parameter values in general and specific values for Pennsylvania when appropriate.

Absorption (bioavailability) factors are thoroughly covered. The recent publication by Ryan et al.[26a] on plant uptake of contaminants is utilized. The report covers 15 exposure pathways. It also covers (perhaps unnecessarily) loss of soil by erosion.

Since the RAFT methodology (unlike AERIS or PPLV) considers contaminant transport, it could provide much more precise estimates of exposure over time.

The model is open to review and modifications, given its spreadsheet format. Other than equations for estimations, it makes few assumptions about the site. Therefore, it is excellent for an experienced user. In addition, the models are finite source or mass conserving models. They are likely to provide more realistic estimates of actual exposure.

Weaknesses

There are several disadvantages to this system that make it less desirable than the AERIS system. In the first place, there is no specific consideration of metals. All the equations relating to transport from soil into air, or from soil into plants, pertain only to organic compounds. Thus, lead, which is a contaminant of petroleum waste and not an uncommon contaminant at wood treatment facilities, would not be adequately addressed with this method.

RAFT requires a substantial amount of data input. Additionally, the exposure part of the methodology seems to have been much more carefully researched than the biological/risk assessment side. In general, the methodology defaults to assumptions about contact, absorption, and bioavailability previously published by the Environmental Protection Agency. An assessment of the conservatism of these assumptions seems warranted.

For example, some of the default assumptions for exposure may be too conservative. The soil ingestion assumptions do not consider the most recent work of Calabrese.[4a] Sediment ingestion factors are provided which essentially assume an ingestion rate similar to that of soil. There is no basis provided for this assumption and, considering the high water content of sediments, it is unlikely that sediments would adhere to the hand to the same extent as soil or housedust. The vegetable consumption factors also appear high, although that may be compensated for by the moderate exposure frequency assumption. In any case, all the exposure assumptions described in this method should be re-evaluated.

A major weakness of the RAFT approach is that it does not take an integrated approach to risk calculations. Specifically, one cannot input target risk levels and derive an acceptable cleanup level for all exposure routes. The RAFT system essentially works in the reverse direction, starting with concentrations of contaminants in environmental media and then calculating risk on an individual medium basis. Total risk is then calculated by summing all the risks from the several media. An approach which is capable of going in both directions would be very desirable.

Some phenomena are overly simplified. For example, the loss of chemicals as vapors from the unsaturated and saturated zones is more complex than indicated.

The biologic half-life in water (surface and ground) seemed to be neglected, and the full text of formulae need to be presented.

The lack of a relationship between concentration in soil and in an intermediate receptor (in particular mother's milk) precludes calculation of cleanup criteria. Forward calculation is really only beneficial for endangerment assessment. The model was not designed for petroleum product mixtures.

There are currently no surface water fate and transport algorithms, which could also provide a means to estimate fish concentrations for this ingestion option.

The exposure values of the risk estimation algorithms (one algorithm for carcinogenic and one for noncarcinogenic) allow the user to supply environmental concentrations or use those that are generated by the fate and transport algorithms. Utilizing the algorithm generated values is a desirable option. However, there are several endpoints which are not estimated previously, and the user must derive an endpoint. These include exposure to contaminants due to: ingestion of water (from surface water — the model is not explicit about the difference between groundwater and surface water), fish, meat, and indoor household dust, and inhalation of contaminated outdoor particulates. Two comments on this: (1) the manual does not make this clear to the user — he or she is left alone to figure out which endpoints are estimated previously and which must be generated uniquely, and (2) there are analogous simplistic models that could estimate surface water concentrations, resulting concentrations in fish, concentrations in beef that could arise if the beef graze over a contaminated site, and outdoor particulate concentrations from windblown contaminated soil. The developers of this model may wish to consider adding these additional algorithms to their model.

Exposure assessment for humans from concentrations in soil, air, and water are based on assumptions which are somewhat at variance with assumptions used by others. Some uncertainty is inherent in all of these assumptions and which of them is most representative is debatable.

This document addresses fate and transport as well as human exposure and considers depletion of the chemical through runoff and evaporation. However, bacterial degradation is not considered. The assumptions for plant uptake seem rather simplistic and metabolism of chemicals by plants and evaporation from leaves are not considered. The result is that the persistence of the chemicals may be overestimated. It is also not clear whether and how the effect of dispersion, weather, and climate, in general, are considered. Furthermore, it does not appear that this model has been validated with real data. The uncertainties are not stated.

Comments

Overall, the RAFT approach is well described and easy to understand. However, it lacks the graphic capability of the AERIS method and does not calculate cleanup levels, but rather risk levels starting with concentrations in individual media. Exposure assumptions are sometimes inappropriate. RAFT lacks the versatility and user-friendly quality of the AERIS model.

The input data requirements seem to be already available for petroleum compounds but are not provided.

Two key sections are under development, but are currently not in the version of the manual that is being reviewed here. These need to be completed.

The manual should be more realistic in the introductory version of the manual that is being reviewed here.

To ensure that the RAFT system is used properly, there should be an accompanying report that discusses in greater detail the mathematical formulations, assumptions, and supporting data for each of the modules. Efforts to validate the various modules should be made. No attempts are made to treat uncertainties or sensitivities in the parameter outputs due to uncertainties in inputs.

An automatic "restrictor" on the input parameters could minimize the chance of generating an unreasonable result.

The "VEGGIES" model should have a link to the soil depletion models in order to calculate a time-varying soil water concentration for plant uptake.

Risk Assistant — Overview of Microcomputer Software to Facilitate Assessment of Hazardous Waste Sites[27] (May 1989)

Developer

The Hampshire Research Institute
1800 Diagonal Road, Suite 150
Alexandria, VA 22314

Available from	*Phone*
The Hampshire Research Institute Contact: Stephanie Murphy 1800 Diagonal Road, Suite 150 Alexandria, VA 22314	703/683-6695

Software and Manual available June, 1991.

Description

Risk Assistant serves as a vehicle for the retrieval of toxicological information on carcinogenic and non-carcinogenic chemicals through various databases and other reference sources. It addition, it reports on and incorporates current regulatory standards and guidelines in modeling the environmental fate and transport of contaminants and conducting exposure assessments.

The environmental fate system contains three programs that are designed to assist the user in predicting environmental concentrations pertinent to the exposure assessment system: environmental distribution, transport model selection assistant, and environmental transport model.

The program assumes that the user can supply the contaminant concentration at the point of exposure by either running models available outside of Risk Assistant or on the basis of site-specific monitoring. Further, it assumes this concentration will remain constant throughout the exposure period, provides defaults for the exposure conditions (i.e., contact rate, body weight, exposure duration, etc.), and calculates the exposure and risk.

Health Evaluation

Basis in Science

A database of 300 chemicals is included in the program containing toxicity constants (cancer potencies and reference doses from IRIS), basic chemical properties, and environmental standards. The following exposure scenarios are included: water ingestion, fish ingestion, air inhalation, soil ingestion, vegetable ingestion, and beef/milk ingestion. Cancer risks are added across chemicals for the same exposure pathway, but not across pathways. Hazard indices are not added across chemicals or pathways.

Additionally, the program contains a "Quick Risk" option where the user specifies the chemical and concentration and the program automatically calculates the exposure/risk for the relevant exposure scenarios. The user cannot control selection of scenarios or parameter values.

Reports are automatically generated listing parameter assumptions, citations, and sources of uncertainty. Help screens are available to assist the user in deciding whether to accept defaults.

Although various scenarios are said to be addressed, this report lists only six, and provides no pathways or equations. It would appear (page 14) that Mackay's fugacity equations are the basis for estimating environmental distribution, but a choice of transport models is also presented. The system includes unspecified chemical property data (but not estimation methods), toxicity data (IRIS), and chemical synonym and regulatory standard information bases. The overall process is quantitative risk assessment.

Applicability

The Risk Assistant software can be applied to multiple sites with a variety of contaminants. Emphasis is placed on human health effects of individual chemicals (despite inclusion of water quality standards) at existing exposure levels — i.e., baseline risk assessment or "endangerment assessment"; there is evidently no direct provision for calculating cleanup levels. The methodology is not petroleum product- or mixture-oriented.

The report does not generate soil cleanup levels. Where transport is not an issue, the report could be used on a trial and error basis to check risks associated with various soil levels. However, this would only reflect soil ingestion, plant uptake, and beef uptake. If the exposure pathways were relevant, it would not be realistic.

The procedures are potentially applicable for computing risks at a wide variety of sites where the user can measure or model (outside of this program) the environmental concentrations and they can be assumed to remain constant over the exposure period.

The Risk Assistant software has the capability of being made site- and chemical-specific.

Although this model is more chemical-specific, as opposed to site-specific, one can provide basic information about a specific site and information on the location and the levels at which the concentrations of the hazardous chemicals were detected.

Site-specific parameters involving the exposed population (i.e., contact rates) and exposure durations can be changed. Physical parameters affecting transport, such as wind speed, depth to groundwater, etc., are not used in the program and therefore cannot be incorporated.

Information from the databases and the model's simplifying assumptions may be modified or augmented as required for the particular site.

Multimedia Relationships:

The software focuses on exposure through water. Risk Assistant predicts how chemicals might be transported from a waste site through the environment and transferred between different environmental media (e.g., air, water, soil, and sediments).

This program generally does not contain transport equations. However, a few very simple media transfer equations are included. The indoor air concentration of volatiles contained in domestic water are calculated using a one-compartment mixing model and assumptions for volatilization fraction, water use, and ventilation rate. Beef/milk concentrations are calculated using a distribution ratio between the soil and beef/milk.

Attention to equilibria between air, soil, and sediment, as well as groundwater transport, is indicated. It does not seem that vaporization from soil is included.

Input Data Requirements

Data bases are provided, but may be inadequate for some sites. Site-specific data would be supplied by the user. The only obligatory input data that the user must supply is the contaminant concentrations in the media of concern. Additionally the user can change the defaults for parameters relating to behavior of the exposed population, i.e., contact rates and durations.

Chemical-specific default values (e.g., half-life) and site-specific information is needed (e.g., contaminant location and soil porosity).

Strengths

Risk Assistant is computerized and user friendly and is applicable to the evaluation of several contaminated soil situations.

Standardized exposure values included in program database can be modified.

Risk Assistant offers ease and speed for deriving preliminary risk estimates, an extensive database of toxicity constants, and automatic report generating.

Weaknesses

The focus is on individual compounds and not on complex mixtures.

The IRIS database is used to derive all of the constant values the program uses. This is a nonpeer reviewed database. Recent outside reviews have shown numerous errors. The regulatory standards portion of the package contains only water quality standards.

The model does not allow for use of carcinogenic potency factors derived by different models.

The model does not allow for use of reference doses derived independent of IRIS or EPA.

The model does not appear to be as flexible or able to handle as extensive a set of exposure scenarios as the PPLV approach and offers a limited number of exposure scenarios.

The model is not designed for petroleum product mixtures. No provision is made for estimating chemical properties or toxicity criteria when they are not in the databases.

The user has no way of evaluating (from this document) the equations to be applied.

The focus is on baseline risk assessment ("endangerment"), not on cleanup.

Bioavailability is apparently not considered.

The presentation does not state which pathways (e.g., via consumption of vegetables) are included.

This model does not take into account exposure to hazardous contaminants via the drinking water (i.e., groundwater contamination) and surface water contamination.

No statistical methods are available for the calculation of dermal exposures.

This model does not take into account interactive effects (i.e., mixtures) between soil contaminants.

This model, as with other models reviewed, does not take into account the probability or improbability of site access (i.e., barbed-wire fence, 24-hour security). This type of information is significant when taking a site-specific approach.

Lack of transport equations could be considered a weakness, although these are generally available elsewhere. Currently no easy way exists to import results of such models in Risk Assistant. Current interface lacks user friendliness. A new version is scheduled for release and may greatly improve this feature.

Comments

Based on the overview manual, it was difficult to determine what, if any, advantages this system had over previously reviewed methodologies. Information on regulatory standards, guidelines and toxicity (as listed in IRIS) is all easily available elsewhere. (IRIS for example is a database accessible via computer dial-up). The available environmental fate models were not described, but would presumably be similar to other easily available programs such as GEMS. The utility of the "Exposure Assessment" program could not be directly assessed since the various exposure scenarios were not identified and the assumptions were not listed. It seemed that the "Exposure Assessment" and "Risk Characterization" programs were essentially on the line version of the EPA Superfund manual. If this is so, this part of the Risk Assistant program would be similar to RAFT (although it would probably be less sophisticated).

This documentation is inadequate for CHESS evaluation of the software it is intended to describe.

Risk Assessment Guidance for Superfund Vol. I — Human Health Evaluation Manual (Part A) — Interim Final EPA/540/1-89/002 December 1989[28]

Developer

U.S. Environmental Protection Agency
Office of Emergency and Remedial Response
Washington, DC

Available from

National Technical Information Service (NTIS)

Order Number: 540/1-86-060

Phone

703/487-4650

General Information

703/487-4600

Description

The Superfund manual is the compilation of several years of evaluating the human health risks posed by exposure or potential exposure to contaminants at Superfund sites. The derivation of risk values are based on "classic" risk assessment principles and methodologies. The basis of this manual is to determine if any mitigative/remedial action should be taken to prevent exposures to hazardous chemicals and thus protect human health.

Since the publication of the *Superfund Public Health Evaluation Manual* (SPHEM, October 1986), the significance of health outcome information and general human health concerns have become a priority through the mission and various program areas of the Agency for Toxic Substances and Disease Registry (ATSDR).

The significance of ATSDR's qualitative, site-specific approach is narrated within this current manual.

Thus, despite the uncertainties identified with quantitative chemical-specific risk assessments, the significance of qualitative, site-specific health assessments are identified and defined. EPA's determination of remedial alternatives that would be best protective of public health is complemented by the ATSDR's evaluation of these remedial alternatives and the possible health follow-up (e.g., pilot health studies, epidemiological studies, exposure/disease registries).

EPA's Superfund Human Health Assessment Manual 3 proceeds from a relatively simple qualitative assessment of available information to a more detailed quantitative risk assessment. The public health evaluation involves three major components:

- baseline site evaluations
- public health assessment of the no-action alternative
- development of design goals and estimation of risk for remedial alternatives

Health Evaluation

Basis In Science

The manual presents broad recommendations for assessing risks associated with exposure to contamination at Superfund sites. Current and future land uses are defined and pathways are identified. Both data and modeling are utilized. (Since Part A concerns baseline risk assessment, data are emphasized, rather than modeling, as compared to what one would see for a feasibility study risk assessment). Site-specific data are required as input. The more complex scientific approaches to exposure assessment are not presented, but the user is referred to other documents. The overall process is quantitative risk assessment. The Superfund Manual has a good basis in science, but is not as well-referenced as possible.

Applicability

The manual presents a site-specific approach for assessing risks associated with exposure to contaminated sites. Emphasis is on human health effects of individual chemicals (environmental effects are treated in Volume II), and carcinogenic risks and hazard indices are summed. Concern is primarily with on-site contamination.

While petroleum contaminated soils are not specifically addressed, an approach for assessing risks associated with exposure to contaminated soils is presented.

The information in this manual is applicable to a variety of sites and land uses (e.g., residential, commercial, industrial, recreational, and agricultural). In order to assess the probability/possibility of exposure, one must be knowledgeable about several factors (e.g., site access including 24-hour security and fencing with barbed wire, nonvegetated vs. vegetated cover, wind flow, and direction). Most of this information is obtained during the site visit and the preliminary investigation conducted by the EPA. Contaminants from a variety of classes (e.g., organics, inorganics, and radioactive materials) are evaluated for health concerns based on acute and chronic risk values. These values are based on several regulatory guidelines and the toxicity information is obtained from a variety of sources (e.g., IRIS, ATSDR toxicological profiles, and TOXNET/TOXLINE).

The approach is quite site-specific; instructions are detailed. The approach could be used to develop a site-specific assessment by a knowledgeable assessor; however, it is not a useful tool for quickly assessing contaminated sites.

This manual does take into consideration various site-specific characteristics (e.g., soil characterization including soil porosity, demographics, e.g., location of nearest residence, age, worker population, and susceptible populations including women of childbearing age, children, and the elderly). This type of information is obtained during the preliminary investigation of the site and the workplan. Site-specific information can be obtained directly from the site PRP, as well as through a site visit. Site-specificity is incorporated into the exposure calculations by basing the assessment on samples taken from the site (or areas impacted by the site). Other site-specific characteristics can be incorporated depending on the models used to evaluate fate and transport (no specific models are recommended in the manual). Characteristics of the population in the vicinity of the site can also be incorporated into the equations for estimating exposure. ATSDR staff conduct site visits on all Superfund sites as an essential part of the health assessment process.

Multimedia relationships

There is very poor treatment of indirect pathways of exposure. No attempt is made to derive exposure concentrations. Instead, frequent reference is made to other sources, in particular to the Superfund Exposure Assessment Manual. The discussion of exposure assessment relies heavily on as many as 12 other EPA reference manuals.

Multimedia relationships are considered in two major ways. First, they are considered in any fate and transport modeling that is conducted to calculate concentrations in environmental media. Second, exposure is estimated for each exposure pathway separately. The risks to populations that may be exposed by more than one pathway are assessed by combining the exposure estimates for the appropriate pathways. In this way, exposure to more than one medium can be considered.

The multimedia relationships are discussed and reasonably complete. Environmental fate and transport of contaminants in the various media are discussed and the role of various physical and chemical characteristics (e.g., K_{oc}) are taken into account. Potential human exposure pathways are discussed in some detail, except for the inhalation of particulates.

Input Data Requirements

No automated data handling is provided by the EPA in this regard. Numerous equations and some receptor-specific values are provided. Properties of the contaminants must be obtained elsewhere (in contrast to the old Superfund Public Health Evaluation Manual). Guidance towards some good sources is provided. Virtually nothing is said about the estimation of physico-chemical or toxicological properties. There is no accompanying computer software package.

Data requirements include estimates of the concentration of chemical present in the site media (obtained from site-specific measurements or modeling), estimates of population characteristics that influence exposure (usually obtained from already compiled sources or by using best professional judgment), route-specific toxicity values for each chemical of interest, ingestion rate (200 mg/d for children ages one through six years; 100 mg/d for those greater than six years of age), and body weight. Surface area would be taken into account as an input parameter for dermal contact with contaminated soil.

Strengths

The manual is very complete; it is a very good resource offering an orderly presentation of all the pertinent issues and is easy to read and understand. It is an excellent primer on risk assessment and the example problems in Chapter 6 are useful. The approach is site-specific and flexible and it yields a conservative estimate of risk.

The approach can be run "backwards" to calculate cleanup levels. This document provides an excellent overview of the risk assessment process, from scoping through data acquisition and application, to documentation and review. The guidance is generally insightful and on target. The graphics are well done and serve as good examples for risk assessment document formatting. The policy of suggesting sources of information on contaminants, rather than providing that information, has considerable merit; the sources (such as IRIS) are more likely than a manual such as this to be up to date. This document is the authoritative voice of the U.S. EPA and is very good; it has corrected some of the major deficiencies of its predecessor.

The primary strengths of this manual revolve around the allowance for site-specificity through information obtained from preliminary investigations, work plans, remedial investigations, and feasibility studies. Health outcome information may be utilized in determining remedial action and any additional health study follow-up. This manual illustrates effective integration of quantitative and qualitative human health evaluations from exposure to contaminants at Superfund sites.

Weaknesses

This manual is not an easy-to-use tool for addressing small or large contaminated sites or for assessing a petroleum contaminated site. If a large site were contaminated, the document may well be useful, but it would be a cumbersome journey. Also, there is no associated computer software available.

Risks from exposure to mixtures are estimated by assuming additivity. This may also be considered a strength.

Toxicity values for all chemicals of concern are not always available. The uncertainty in risk estimates (at least an order of magnitude) is much greater than the resolution desirable for cleanup decisions (where a factor of two or three can make a big difference).

The major disadvantage of this manual is that there is still a prevalence in uncertainties with the utilization of exposure duration values as well as other "guesstimate values". In addition, uncertainties in evaluating chemical interactions or chemical mixtures is also still a significant data gap in the evaluation of exposure risks from a litany of contaminants at Superfund sites. The use of reference doses derived for one exposure pathway for assessing risk from exposure via another pathway may be inappropriate. For example, the use of oral RfDs for assessing risk from dermal exposure may be inappropriate. Lack of this information represents a significant data gap.

This document covers the toxicology of mixtures well, but not the physical chemistry; it is not directly concerned with petroleum products.

The concatenation of conservatively biased values, especially when coupled with extremely low upper-bound risk estimates for carcinogens, fosters the illusion of far greater human risks than must actually be occurring. Moreover, weight-of-evidence for carcinogens is not utilized quantitatively. The concept of "Reasonable Maximum Exposure" leads to treatment of an entire population as if it were likely to suffer at the maximum level. Thus, of a million people in a given area, a very few might be exposed, for example, to an upper-bound estimated risk of 10^{-4}, whereas most could be exposed to a far lower risk; yet the risk assessment would not indicate the nature of the distribution, only the highest value. The manual (paragraph 7.7.1) seems to wrongly imply that toxic effects occur at criterion levels; since RfDs are not toxicity values, the wording should be changed.

The manual's exposure assessment sections (especially 6.4 to 6.6) are weak and incomplete because they rely heavily on references to the *Superfund Exposure Assessment Manual* (SEAM); unfortunately, SEAM is not as good as it might be. Thus, the exposure assessment guidance offered by the manual is incomplete and the user is given inadequate support. Some pathways, such as the food chain, could easily be overlooked.

The derivation of NOAELs from LOAELs, described in the predecessor *Superfund Public Health Evaluation Manual* (Appendix D) which uses severity ratings, has unfortunately been eliminated; this was a worthwhile concept that still has a useful place.

Another useful tool that is not found in this manual is the derivation of reference doses from acute doses (as developed by Layton et al.[28a]). The term "developmental toxicants" is applied to such a broad range of effects that one might apply fetotoxic effects numbers to the protection of adolescent populations; the appropriate caveats need to be introduced.

One accepts that the manual is not complete, and that it is concerned only with baseline risk assessment, with cleanup criteria to follow as part of the risk assessment connected with feasibility studies. It is not quite clear, however, whether a baseline risk assessment, with appropriate modeling, is supposed to consider future land uses under the no-action option. The document does mention future populations and land uses briefly, but with no great enthusiasm. One may accept the strong warning to avoid intrusion into risk management as regards a baseline risk assessment, but the manual should concede that consideration of risk management alternatives is essential to establishing cleanup criteria (i.e., conducting risk assessments in connection with feasibility studies).

This document will soon represent a *de facto* standard for assessing the health risks of toxic substances at contaminated sites because of its role as a key source of guidance in Superfund assessments. Chapter 6 is the most important chapter in relation to the CHESS review. It includes several equations for calculating water-, soil-, and air-based exposures to contaminants. Also presented are parameter values for the equations and supporting references.

Although the document provides median and upper-bound (e.g., 90th and 95th percentile) clauses for many of the parameters, it does not discuss the statistical techniques for propagating uncertainties in multiplicative equations (except for some obtuse references in Chapter 8). In fact, if you follow the approach presented (i.e., selecting upper-bound parameter values for a health conservative calculation), no statement can be made as to how conservative the estimated exposure is.

One other issue related to the suggested input parameters is an obvious lack of peer-reviewed publications supporting the recommended values. It appears that some of the input values listed are much too high, even for the so-called "average values". Examples include the inhalation and drinking water rates. In summary, while the basic exposure equations appear correct, no methodology is presented for using them in a way that provides results that can be associated with a given level of uncertainty (e.g., upper 95th percentile of exposure).

Comments

The manual is not acceptable as a "ready tool" for a state or local agency but it does serve as a good reference manual. For the current purposes of the CHESS Committee, it is an excellent document for identifying issues that need to be considered. In general, this document is very well-organized, "reader friendly", and very nicely integrates the basic principles of both risk and health assessments.

REFERENCES

1. "The Development of Soils Cleanup Criteria in Canada," Volume 2 — Interim Report on the "Demonstration" Version of the AERIS Model, Senes Consultants, prepared for the Decommissioning Steering Committee 1988-12-15.
2. "Contaminated Soil Cleanup in Canada." Volume 5 — Development of the AERIS Model, Final Report, Senes Consultants, prepared for the Decommissioning Steering Committee (September 1989).
3. "Contaminated Soil Cleanup in Canada." Volume 6 — User's Guide for the AERIS Model, Senes Consultants, prepared for the Decommissioning Steering Committee (September 1989)
4. Ibbotson, B. G. et al. "A site-specific approach for the development of soil cleanup guidelines for trace compounds," *Petroleum Contaminated Soils, Remediation Techniques, Environmental Fate and Risk Assessment*, Volume 1 (Michigan: Lewis Publishers, 1989) pp. 321–341

4a. Calabrese, E. J., R. Barnes, E. J. Stanck III, H. Pastides, C. Gilbert, P. Veneman, X. Wang, A. Lasztity, and P. Kostecki. "How Much Soil Do Young Children Ingest: An Epidemiologic Study," *Reg. Toxical. Pharmacol.*, 10:123–137 (1989).

5. Wagner, J., and M. Bonazountas. "Potential Fate of Buried Halogenated Solvents via SESOIL." ADL for U.S. Environmental Protection Agency, Office of Toxic Substances (January 1983).
6. Watson, D., and S. Brown. "Testing and Evaluation of the SESOIL Model," Anderson-Nichols & Co, prepared for U.S. Environmental Protection Agency, Environmental Research Lab. Athens, GA (August 1985).
7. Bonazountas, M. and J. Wagner. " 'Sesoil,' A Seasonal Compartment Model," ADL and DIS/ADLPIPE, prepared for U.S. Environmental Protection Agency, Office of Toxic Substances (May 1984).

7a. Hetrick, D. M., C. C. Travis, S. K. Leonard, R. S. Kinerson, "Qualitative Validation of Pollutant Transport Components of an Unsaturated Soil Zone Model, (SESOIL)," Oak Ridge National Laboratory/TM–10672, March 1989.

8. "Personal Computer Version of the Graphical Exposure Modeling System," User's Guide, General Sciences Corp, prepared for USEPA/OTS Contract #68024281 (September 1989).
9. Leu, D. "California Site Mitigation Decision Tree Manual," Department of Health Services, Toxic Substances Control Division (May 1986).
10. "Leaking Underground Storage Tank Manual: Guidelines for Site Assessment. Cleanup, and Underground Storage Tank Closure," LUFT Task Force, State of California, Sacramento, CA (May 1988).
11. McKone, T. E., L. B. Gratt, M. J. Lyon, and B. W. Perry. "Geotox — User's Guide and Supplement," Lawrence Livermore Na-

tional Laboratory/U.S. Army Medical Research and Development Command, Project Order 83PP3818 (May 1987).

12. McKone, T. E., and D. W. Layton. "Screening the Potential Risks of Toxic Substances Using a Multimedia Compartment Model: Estimation of Human Exposures," *Reg. Toxicol. Pharmacol.*, 6:359–380 (1986).

13. McKone, T. E. and D. W. Layton. "Exposure and Risk Assessment of Toxic Waste in a Multimedia Context," Lawrence Livermore National Laboratory, presented at Air Pollution Control Association, (May 1986).

14. McKone, T. E. "Geotox — Simulating Contaminant Behavior and Human Exposure," *Energy Technol. Rev.* pp. 14–20. (May 1987).

15. Stokman, S. and R. Dime. "Soil Cleanup Criteria for Selected Petroleum Products," *J. Risk Assess.*, (1986) 342–345.

16. Hawley, J. K. "Assessment of Health Risk from Exposure to Contaminated Soil," *Risk Anal.*, 5(4) (1985).

17. "Draft Interim Guidance for Disposal Site Risk Characterization — In Support of the Massachusetts Contingency Plan," Massachusetts Department of Environmental Quality Engineering, Office of Research and Standards (October 3, 1988).

18. Tetra Tech, Inc. "MYGRT: An IBM Personal Computer Code for Simulating Solute Migration in Groundwater," Electric Power Research Institute, Palo Alto, CA. Project RP2485-1, May 1986.

19. Enfield, C., R. Carsel, S. Cohen, T. Phan and D. Walters. "Approximating Pollutant Transport to Groundwater," *Ground Water* 20:(6) (1982).

20. Carsel, R., C. Smith, L. Mulkey, J. Dean and P. Jowise. "User's Manual for the Pesticide Root Zone Model (PRZM)," U.S. EPA Environmental Research Lab, Athens, GA (December 1984).

21. Dean, J. and R. Carsel. "Agricultural Chemical Use," Woodward-Clyde Consultants, prepared for U.S. EPA, Athens, GA (1988).

22. Brown, S. and S. Boutwell. "Chemical Spill Exposure Assessment Methodology," CH2M HILL, prepared for Electric Power Research Institute, (January 1988).

23. Brown, S. and A. Silvers. "Chemical Spill Exposure Assessment," *Risk Anal.* 6:(3) (1986).

24. Small, M. "The Preliminary Pollutant Limit Value User's Manual," U.S. Army Biomedical Research and Development Laboratory, Ft. Detrick, Frederick, MD, Technical Report 8918 (July 1988).

25. Rosenblatt, D. H., J. C. Dacre, and D. R. Cogley, "An Environmental Fate Model Leading to Preliminary Pollutant Limit Values for Human Health Effects," *Environmental Risk Analysis for Chemicals*, R. A. Conway, Ed., (New York: Van Nostrand Reinhold Co., 1981).

25a. Hawley, J. K., "Assessment of Health Risk from Exposure to Contaminant Soil," *Risk Anal.* (5)4:289–302 (1985).

26. "RAFT — User's Manual for Risk Assessment/Fate and Transport

(RAFT) Modeling System," Scientific Services Section of the Pennsylvania Bureau of Waste Management (1989).

26a. Ryan, J. A., R. M. Bell, J. M. Davidson, and G. A. O'Conner, "Plant Uptake of Non-Ionic Organic Compounds from Soil," 17:2299–2323 (1988).

27. "Risk Assistant — Overview of Microcomputer Software to Facilitate Assessments of Hazardous Waste Sites," Hampshire Research Institute for the U.S. EPA and NJDEP (May 1989).

28. Layton, D. W., D. J. Mallor, D. H. Rosenblatt, and M. J. Small, "Deriving Allowable Daily Intakes for Systemic Toxicants Lacking Chronic Toxicity Data," *Reg. Toxicol. Pharmacol.*, 7:96–112 (1987).

BIBLIOGRAPHY

Baehr, A. "Selective Transport of Hydrocarbons in the Unsaturated Zone due to Aqueous and Vapor Phase Partitioning," *Water Resour. Res.* 23(10):1926–1938 (1987).

Baehr, A., G. Hoag, and M. Marley. "Removing Volatile Contaminants from the Unsaturated Zone by Inducing Advective Air-Phase Transport," *J. Contam. Hydrol.* (4):1–26(1989).

Baehr, A., and M. Corapciouglu. "A Compositional Multiphase Model for Groundwater Contamination by Petroleum Products," *Water Resour. Res.* 23(1):191–213 (1987).

"Contaminated Soil Cleanup in Canada." Volume 5 — Development of the AERIS Model, Final Report, Senes Consultants, prepared for the Decommissioning Steering Committee (September 1989).

"Contaminated Soil Cleanup in Canada." Volume 6 — User's Guide for the AERIS Model, Senes Consultants, prepared for the Decommissioning Steering Committee (September 1989).

"Development of Advisory Levels of PCBs Cleanup," EPA-600/6-86/002 (Appendices A and B describe volatilization models) (1986c).

"Development of Statistical Distributions or Ranges of Standard Factors used in Exposure Assessments," EPA-600/8- 85/010 (1985).

Environmental Protection Agency, Office of Underground Storage Tanks. "Cleanup of Releases from Petroleum USTS: Selected Technologies," EPA/530/-UST-88/001 (April 1988).

ESNR Consulting and Engineering. "Risk Assessment for the Proposed Wheelabrator Falls, Inc. Recycling and Energy Recovery Facility," Prepared for RUST International Inc. (June 1989).

Foskett, W. "Soil Treatment Advisor," Part of EPA/OUST's Corrective Action Triage Software (CATS).

"Guide to the Assessment and Remediation of Underground Petroleum Releases," American Petroleum Institute Publication No. 1628, Washington, D.C.

ICF-Clement. "Environmental Health Risks Associated with Exposure to soils contaminated with polychorinated biphenyls at the Paoli, Pennsylvania site" (1988c).

Israelsen, C. D., C. G. Clyde, J. E. Fletcher, E. K. Israelsen, F. W. Haws, P. E. Packer, and E. E. Farmer. "Erosion Control During Highway Construction: manual on Principals and Practices," National Cooperative Highway Research Program, Washington, D. C. (1980).

Kostecki, P.T., E.J. Calabrese and H. Horton. "Review of Present Risk Assessment Models for Petroleum Contaminated Soils," in *Petroleum Contaminated Soils,* P.T. Kostecki and E.J. Calabrese, Eds. (Chelsea, MI: Lewis Publishers, Inc., 1988).

Lyman, W. J., W. F. Reehl and D. H. Rosenblatt. *Handbook of Chemical Property Estimation Methods.* (New York, NY:McGraw-Hill Book Co., 1982).

"Rapid Assessment of Potential Groundwater contamination under emergency response conditions," EPA-600/8–83/030 (1983).

"Risk of Unsaturated / Saturated Transport and Transformation of Chemical Concentraton, Vol 1: Theory and Code Verification," EPA-600/3-89/048a (1989).

"Superfund Exposure Assessment Manual," EPA(1986b).

"Superfund Public Health Evaluation Manual," EPA- 540/1-86/060 (1986a).

"The Development of Soils Cleanup Criteria in Canada," Volume 2 — Interim Report on the "Demonstration" Version of the AERIS Model, Senes Consultants, prepared for the Decommissioning Steering Committee 1988–12–15.

U.S. EPA, "Risk Assessment Guidance for Superfund: Volume I — Human Health Evaluation Manual (Part A)," Interim Final, (December 1989).

INDEX

E

F

G

H

I

N

O

P

Q

R

T

U

V

W

X

Z